Blueprint for a Methanol Society

Junji Nakamura • Masayoshi Ishida

Blueprint for a Methanol Society

Toward Carbon-Neutrality

Springer

Junji Nakamura
International Institute for Carbon-Neutral Energy Research, Kyushu University
Fukuoka, Japan

Masayoshi Ishida
Faculty of Eng., Info. and Systems
University of Tsukuba
Tsukuba, Japan

ISBN 978-981-95-3464-7 ISBN 978-981-95-3465-4 (eBook)
https://doi.org/10.1007/978-981-95-3465-4

The original submitted manuscript has been translated into English. The translation was done using artificial intelligence. A subsequent revision was performed by the author(s) to further refine the work and to ensure that the translation is appropriate concerning content and scientific correctness. It may, however, read stylistically different from a conventional translation.

This Springer imprint is published by the registered company Springer Nature Singapore Pte Ltd.
The registered company address is: 152 Beach Road, #21-01/04 Gateway East, Singapore 189721, Singapore

Preface: Authors' Message and Structure of this Book

What is global warming? What is the energy problem? In modern society, fossil fuels are the primary source of energy, but their combustion inevitably results in the emission of carbon dioxide (CO_2) into the atmosphere. This is the core of the issue: a straightforward yet profound problem. If global warming continues unabated, it threatens not only the survival of humankind but also a vast array of plant and animal life. In effect, the fate of Earth's biosphere rests in human hands.

The impact of global warming is already evident, with increasingly frequent and severe natural disasters—floods, wildfires, and heatwaves—being reported worldwide. These changes are unfolding within a single human generation, which is unprecedented in the Earth's recent climate history. Although living organisms have a degree of adaptability, the limits are clear, and humanity must recognize that our civilization may be accelerating the crisis rather than solving it. Despite this, many aspects of daily life and society continue as if the problem were someone else's responsibility. We are encouraged to use air conditioning to avoid heatstroke while luxury cars with single occupants continue to congest our streets. This apparent complacency is deeply concerning.

Following the 28th Conference of the Parties (COP28) to the UN Framework Convention on Climate Change in 2023, Secretary-General António Guterres declared that "the era of global boiling has arrived," calling for a complete phase-out of fossil fuels. July of that year was officially recognized as the hottest month in recorded human history, and in January 2024, the World Meteorological Organization (WMO) announced that the previous year's global average temperature had risen by 1.45°C compared to the pre-industrial baseline—just 0.05°C below the 1.5°C target established by the 2015 Paris Agreement. The window for meaningful action is rapidly closing.

Despite global commitments to carbon neutrality and the Sustainable Development Goals (SDGs), actual progress has been minimal. As of 2024, global CO_2 emissions exceed 37 billion tons annually, with no sign of significant decline.

The imperative for humanity is clear: we must act now to mitigate global warming to the greatest extent possible. While some argue it may already be too late, we must nevertheless pursue optimal measures with urgency and resolve. The general direction is well known—transitioning to renewable energy—but this alone will not suffice. Of critical importance is the capture of CO_2 at emission sources, especially during the gradual phase-out of fossil fuels. Significant investment is required, and the technology is available. However, large-scale deployment has been hindered primarily by economic and political barriers—such as high upfront costs, insufficient carbon pricing, lack of CO_2 transport infrastructure, and limited public acceptance of carbon capture and storage (CCS) facilities—rather than by technical feasibility.

Captured CO_2 can be converted into synthetic fuels using renewable energy, with methanol being a particularly promising option. Methanol is a versatile energy carrier that can be used in thermal power plants, internal combustion engines, and fuel cells. It also serves as a key feedstock in the chemical industry. A circular carbon economy—CO_2 → methanol → CO_2—is thus both technically achievable and strategically valuable. The concept of a methanol-based energy system has been extensively advocated, notably by Nobel Laureate George Olah, who proposed the "Methanol Economy." However, *practical pathways toward realizing such a society remain underdeveloped.*

This book is structured into three parts:

- **Part I identifies and quantifies the key issues**: global warming, energy systems, challenges of renewable energy adoption, limitations of hydrogen and ammonia, battery constraints, power generation, and industrial emissions.
- **Part II introduces the concept of a methanol society.** It explains why methanol is a viable energy carrier, particularly in capturing CO_2 from thermal power plants, the steel and cement industries, and chemical production. The discussion emphasizes cost, feasibility, and existing technologies.
- **Part III explores pathways for implementation.** It presents strategies involving collaboration among industry, governments, and academia, emphasizing the need for scalable and inclusive solutions that benefit both developed and developing nations. The transition must also support job creation and business development and foster systems-level thinking in the next generation of scientists, engineers, and policymakers.

A distinguishing feature of this book is its quantitative approach. Effective decision-making requires data-driven, numerical analysis. This applies not only to engineers and scientists but also to policymakers and citizens. In particular, the younger generation must engage in informed debate, influence political decisions, and drive responsible consumption. Through education and communication—including social media—we can accelerate the dissemination of accurate information and promote societal change.

This book is co-authored by Masayoshi Ishida, who has long been involved in education and research on electricity and energy systems, and Junji Nakamura, a specialist in methanol synthesis catalysis. For over a decade, the authors have shared a belief in the central role that methanol could play in a future energy system. In the process, they recognized the strong interdependence between the energy and chemical sectors when utilizing renewable energy.

Despite growing evidence supporting methanol's potential, global policy trends have often moved in other directions—many of them lacking rigorous quantitative justification. This book is the result of the authors' conviction that a more rational, data-oriented approach is urgently needed. It is offered as a resource for those seeking to understand and advance the methanol-based energy transition.

While we recognize that some aspects of this book may contain errors or simplifications, we welcome feedback and constructive criticism. Our goal is to contribute to a more sustainable and scientifically grounded discussion of global energy futures. We hope this book serves as a useful step in that direction—and we invite readers to join us in advocating for a methanol-centered societal transformation.

Fukuoka, Japan Junji Nakamura
Tsukuba, Japan Masayoshi Ishida

Contents

Part I
Identifying the Problem

Chapter 1
Global Warming and CO_2 Emissions

Abstract This chapter presents a quantitative analysis of global warming, focusing on the rise in Earth's surface temperature and its primary driver: anthropogenic CO_2 emissions. Scientific consensus—confirmed by over 99% of peer-reviewed studies—shows that human activities, especially fossil fuel combustion, are the dominant cause of recent climate change. Historical data clearly demonstrate the link between energy use and rising atmospheric CO_2 concentrations, particularly since the Industrial Revolution. This chapter explains the greenhouse effect, emphasizing the infrared absorption characteristics of CO_2 and its persistence in the atmosphere, which continues to amplify warming over time. Key environmental consequences include sea-level rise, extreme weather events, and biodiversity loss. This chapter also addresses widespread misconceptions about the physical mechanisms of warming and clarifies the distinction between natural variability and human-induced change. A central theme is that understanding global warming through numerical evidence is essential for policy design and public engagement. By analyzing CO_2 sources, sinks, and long-term climate impacts, this chapter provides a foundational understanding of why urgent action is needed—and why it must be grounded in science.

Keywords Global warming · Anthropogenic CO_2 · Scientific consensus · Greenhouse effect · Climate change · Fossil fuels · Radiative forcing

1.1 Abnormal Rise in Earth's Surface Temperature

Empirical Evidence of Temperature Increases

In this section, we examine empirical evidence indicating that Earth's surface temperature is indeed rising, by comparing two representative datasets. Figure 1.1 shows global temperature data compiled by Our World in Data [1]. Figure 1.2 shows data from NASA's official climate website [2]. These two datasets are consistent

J. Nakamura, M. Ishida, *Blueprint for a Methanol Society*,
https://doi.org/10.1007/978-981-95-3465-4_1

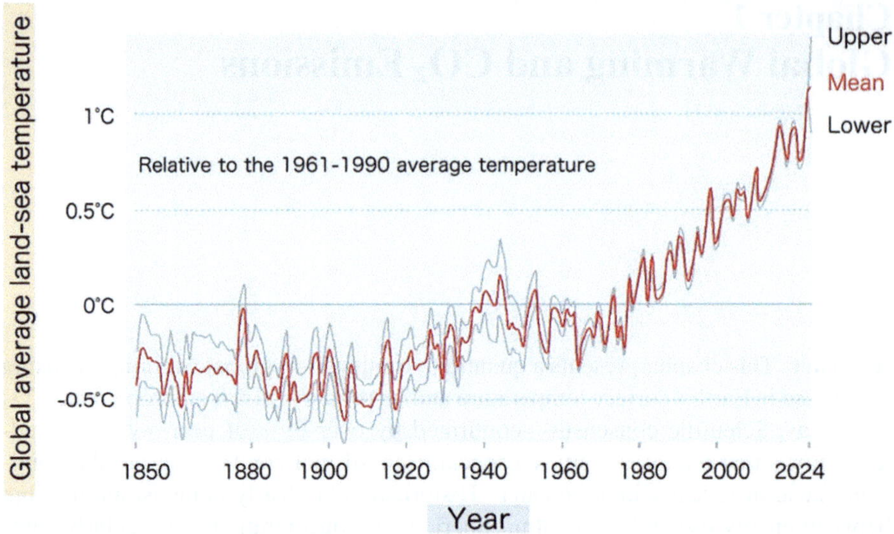

Fig. 1.1 Change of global average temperature (Our World in Data) [1]. Data source; HadCRUT5, (Reprinted from https://ourworldindata.org/climate-change under the Creative Commons Attribution 4.0 International License)

with the results of the 2007 IPCC Fourth Assessment Report [3], which compiled older datasets from multiple sources.

Consistent Trends Across Datasets

Despite differences in methodology and temporal resolution, all three figures reveal consistent trends:

- From approximately 1850 to 1920, global surface temperature remained relatively stable.
- A modest increase was observed between about 1925 and 1945, followed by a period of near-stagnation lasting until around 1975.
- A sustained and significant rise in surface temperature has been evident since approximately 1980.

Thus, the Earth has experienced a continuous warming trend for roughly the past 45 years (1980–2024). As of 2024, global surface temperature is estimated to have increased by approximately 1.2–1.4 °C relative to the pre-industrial baseline (circa 1850). While this level of warming may seem modest, it can have profound consequences for ecosystems and climate stability worldwide.

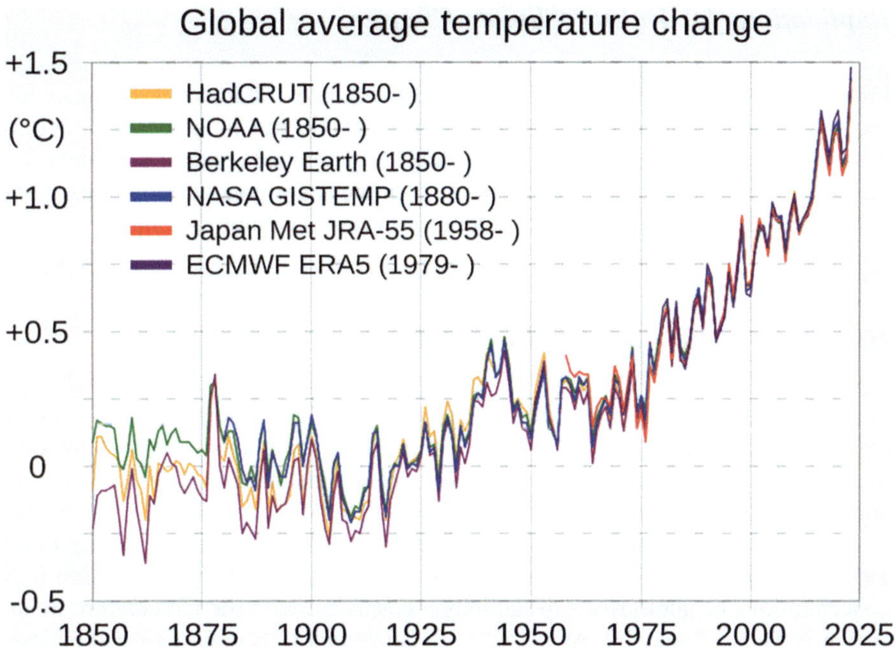

Fig. 1.2 Change of global average temperature [2]. (Reprinted from https://en.wikipedia.org/wiki/File:20200324_Global_average_temperature_-_NASA-GISS_HadCrut_NOAA_Japan_BerkeleyE.svg under Creative Commons Attribution-ShareAlike 4.0 International License)

International Goals and Challenges

The Paris Agreement, adopted at COP21 in December 2015, set a goal of keeping the global average temperature rise "well below 2°C" above pre-industrial levels, while pursuing efforts to limit the increase to 1.5 °C. However, recent observations and projections suggest that achieving the 1.5 °C target may no longer be feasible without immediate and drastic global action.

Insights from the Early Twentieth Century Warming

One point of scientific interest is the warming period from approximately 1925 to 1945, known as the *Early Twentieth Century Warming (ETCW)*. The cause of this warming episode remains partially unresolved. While human activities likely played a role, recent studies attribute ETCW primarily to natural variability, including changes in atmospheric and oceanic circulation patterns, fluctuations in solar irradiance, and reduced volcanic activity. The episode highlights the complexity of the Earth's climate system and the interplay between anthropogenic and natural factors.

Implications for Future Climate Change

From a cautious or even pessimistic perspective, the existence of warming events such as the ETCW suggests that future climate change may be influenced not only by human-induced CO_2 emissions but also by unpredictable natural feedbacks and variability. This underscores the urgency of limiting anthropogenic emissions, as we may face additional warming from natural sources that cannot be controlled.

Historical Trends in Global CO_2 Emissions

Let us now examine the historical trajectory of global carbon dioxide (CO_2) emissions. Figures 1.3 and 1.4 present data on global CO_2 emissions from 1879 to 2022, disaggregated by fuel source—namely coal, oil, and natural gas. These two figures are derived from the same dataset but are divided for visual clarity.

A key observation is the significant increase in CO_2 emissions beginning in the 1950s, which closely corresponds with the period during which atmospheric CO_2 concentrations began to rise substantially—consistent with the temperature trends shown in Figs. 1.1 and 1.2. As of 2022, global annual CO_2 emissions had reached approximately 35 gigatons.

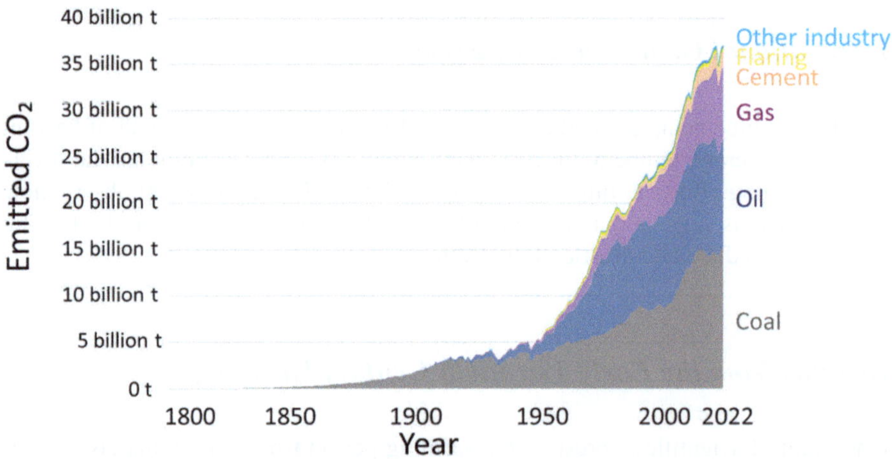

Fig. 1.3 Global CO_2 emissions by fuel type (Our World in Data) [4]. (Reprinted from https://ourworldindata.org/emissions-by-fuel. under the Creative Commons Attribution 4.0 International License)

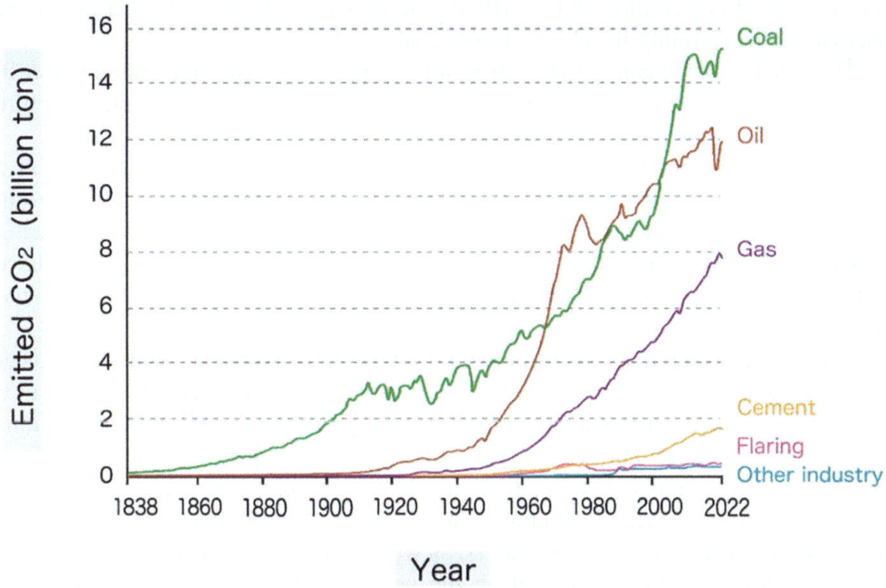

Fig. 1.4 Global CO_2 emissions by fuel type (Our World in Data) [4]. (Reprinted from https://ourworldindata.org/emissions-by-fuel. under the Creative Commons Attribution 4.0 International License)

From Coal to Oil and Natural Gas

To understand this progression in detail, we begin with the period from 1900 to 1950. During this era, global CO_2 emissions averaged around 5 gigatons per year, predominantly driven by coal consumption. Following the Industrial Revolution, coal use expanded rapidly and became the primary energy source for key industrial sectors, including steel production, railways, steamships, and factories. The steel industry, in particular, consumed large quantities of coal for iron smelting and became the backbone of infrastructure, machinery, and transport development.

Coal's influence extended beyond heavy industry. The chemical industry utilized coal-derived by-products such as coke and coal tar, and widespread construction of coal-fired power plants contributed to the expansion of electrification in both industry and residential life. Military demands during World War I (1914–1918) and World War II (1939–1945) further intensified coal consumption.

After World War II, a transition from coal to oil occurred, as illustrated in Fig. 1.4. Oil offered several advantages over coal: higher energy density, ease of transportation, and versatility in refining into a wide array of products. The global proliferation of automobiles was a major factor in the rise of oil consumption. Additionally, oil became an essential feedstock for the petrochemical industry, enabling the production of plastics, synthetic fibers, rubber, and fertilizers. As global infrastructure for oil extraction, refining, and distribution matured, oil became cheaper and more accessible, accelerating its adoption over coal.

This energy shift coincided with rapid economic growth and urbanization in industrialized nations, leading to increased demand for transportation infrastructure and electricity—both heavily reliant on oil.

Beginning around 1970, a third major fossil fuel—natural gas—emerged as a significant contributor to global CO_2 emissions. Technological advancements in exploration and extraction, particularly the development of hydraulic fracturing (fracking) and horizontal drilling, led to a surge in natural gas supply. Extensive pipeline networks were established to transport natural gas efficiently to cities and industrial zones.

Natural gas offered key environmental advantages over coal, including lower CO_2 emissions per unit of energy and higher combustion efficiency. These attributes made it an attractive alternative in response to tightening air pollution regulations and efforts to reduce greenhouse gas emissions. Consequently, many coal-fired power plants were replaced by natural gas plants in several countries.

In addition to electricity generation, natural gas gained importance in industrial applications such as boilers and heaters, as well as residential heating and cooking. It also became a critical feedstock in the chemical industry—particularly in the synthesis of ammonia and methanol, both of which are essential to fertilizers and fuels, respectively.

CO_2 Emissions Tied to the Combustion of Fossil Fuels

What Figs. 1.3 and 1.4 clearly demonstrate is that the rise in global CO_2 emissions is directly tied to the combustion of fossil fuels—coal, oil, and natural gas. These resources have historically served as the dominant energy sources for industrial and economic development (see Sect. 2.1). Therefore, it is evident that any meaningful strategy to reduce CO_2 emissions must begin with a fundamental transformation of the global energy system. In short, an energy revolution is imperative.

Atmospheric CO_2 Concentrations Since 1960

We now examine the atmospheric concentration of carbon dioxide (CO_2) since 1960—an era marked by a rapid rise in anthropogenic CO_2 emissions. Figures 1.5 and 1.6 present two key datasets that reveal both recent and long-term trends in atmospheric CO_2 levels [5, 6].

As shown in Fig. 1.5, the atmospheric concentration of CO_2 has increased significantly from approximately 320 ppm in 1960 to around 420 ppm in 2024, as indicated by the red line. Before the Industrial Revolution, the concentration was

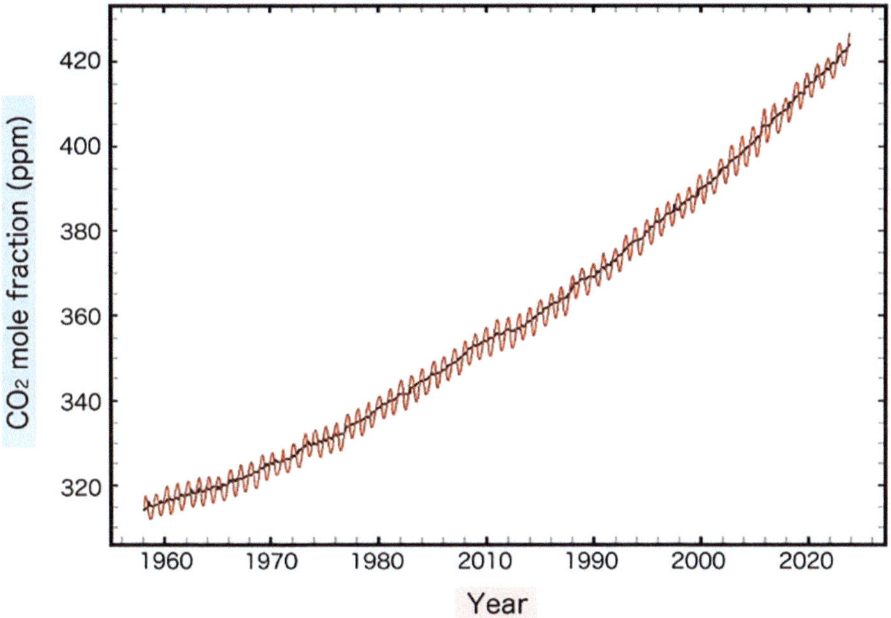

Fig. 1.5 Change of CO_2 concentration in atmosphere measured at Mauna Loa Observatory since 1958 [5]. Figure based on data from NOAA Global Monitoring Laboratory Trends in CO_2. NOAA Global Monitoring Laboratory https://gml.noaa.gov/ccgg/trends/ under Creative Commons Zero v1.0 Universal Public Domain Dedication

about 280 ppm. This sharp increase corresponds directly to the escalation in fossil fuel combustion and other human activities.

CO_2 Levels Over Geological Timescales

Figure 1.6 illustrates changes in atmospheric CO_2 over the past 800,000 years, based on analyses of air bubbles trapped in polar ice cores. These reconstructions reveal that, during glacial periods, CO_2 levels hovered around 200 ppm, while in interglacial (warmer) periods, they reached approximately 280 ppm. Notably, at no point in this 800,000-year record did atmospheric CO_2 levels exceed 300 ppm—until the twentieth century. The current concentration of over 420 ppm is thus anomalously high in the context of natural climate variability and is unprecedented in at least the last 800,000 years.

This rapid rise in CO_2 concentration aligns with the trends observed in global CO_2 emissions (Fig. 1.3) and surface temperature increases (Figs. 1.1 and 1.2).

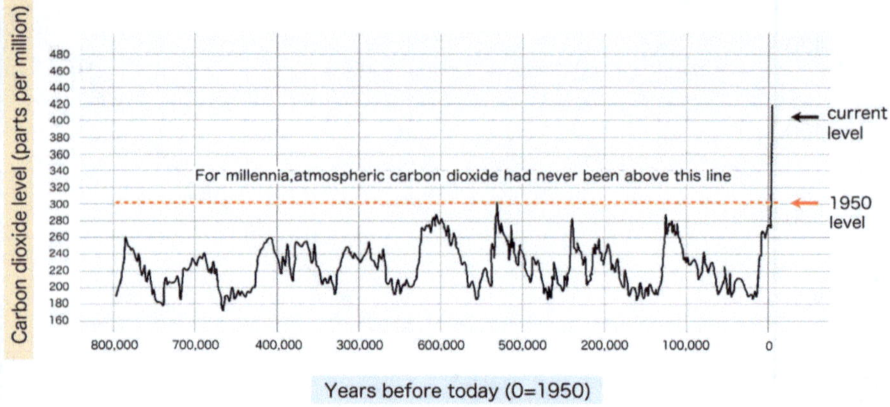

Fig. 1.6 Long span change of CO_2 concentration in atmosphere [6]. Figure based on data from NASA Global Climate Change: Vital Signs of the Planet, Graphic: The relentless rise of carbon dioxide. NASA Science, https://science.nasa.gov/resource/graphic-the-relentless-rise-of-carbon-dioxide/ under Creative Commons Zero v1.0 Universal Public Domain Dedication

Human Emissions and Their Impact

How much CO_2 has humanity emitted? According to Fig. 1.3, cumulative anthropogenic CO_2 emissions from fossil fuel use between 1850 and 2022 amount to roughly 1540 Gt. In addition, land-use changes, such as deforestation, are estimated to have contributed an additional 710 Gt of CO_2. In total, human activities have released approximately 2250 Gt of CO_2 into the atmosphere [7, 8].

Land-use change is a significant and often underappreciated source of anthropogenic CO_2 emissions. Between 1850 and 2022, cumulative CO_2 emissions from land-use change—including deforestation, forest degradation, and peatland drainage (710 $GtCO_2$)—represent nearly one-third of all human-caused CO_2 emissions over that period. A common misconception is that deforestation only contributes to emissions when biomass is burned. In reality, a substantial portion of felled biomass decomposes gradually through microbial processes, releasing CO_2 into the atmosphere over time. Even when wood is harvested for long-lived products, eventual decay or disposal typically results in carbon release. Moreover, the disturbance of forest soils during clearing or conversion to agriculture accelerates the decomposition of soil organic matter, further adding to CO_2 emissions. Therefore, land-use-related emissions must not be neglected when evaluating global carbon budgets or designing mitigation strategies.

Of the CO_2 emitted from the Earth's surface, 40% accumulates in the atmosphere. As of 2024, atmospheric CO_2 concentrations have risen from approximately 280 ppm in the pre-industrial era to over 420 ppm today. This increase of about 140 ppm corresponds to an additional ~900 $GtCO_2$ accumulated in the atmosphere. Given that the total mass of CO_2 currently present in the atmosphere is estimated to be around 3200–3300 $GtCO_2$, this means that approximately 27–28% of the total atmospheric CO_2 has been added by human activity in just the past 170 years.

The remaining portion of the emitted CO_2—roughly 60% of cumulative emissions—has been absorbed by natural systems, including oceans and terrestrial ecosystems, which act as carbon sinks. This redistribution has slowed the rate of atmospheric accumulation but has also altered the chemistry and functioning of these natural reservoirs. Ocean acidification, ecosystem stress, and the potential for carbon sink saturation are growing concerns.

In effect, human activity has not only emitted unprecedented quantities of CO_2 but has also fundamentally disrupted the balance of the global carbon cycle. The atmospheric burden of CO_2 is now significantly greater than at any point in the past 800,000 years, and its long atmospheric residence time ensures that the climate impacts of these emissions will persist for generations. Stabilizing the climate will require not only continued carbon sequestration by natural systems, but above all, a rapid and sustained reduction in anthropogenic CO_2 emissions.

CO_2 Sinks: Oceans, Terrestrial Biosphere, and the Disruption of the Carbon Cycle

Thus far, we have focused primarily on atmospheric CO_2, but to understand the global carbon cycle, we must also consider the **carbon sinks**—natural systems that absorb and store CO_2 from the atmosphere. The two major sinks are oceans and terrestrial ecosystems (especially vegetation and soils). Current estimates suggest that oceans absorb approximately 28% and terrestrial ecosystems absorb roughly 30% of annual anthropogenic CO_2 emissions, playing a critical role in mitigating both atmospheric CO_2 accumulation and the broader impacts of climate change [8, 9].

In the ocean, CO_2 is absorbed through the following series of chemical equilibria:

$$CO_2 + H_2O \rightleftharpoons H_2CO_3 \rightleftharpoons HCO_3^- + H^+ \rightleftharpoons CO_3^{2-} + 2H^+ \quad (1.1)$$

Approximately **90%** of the absorbed CO_2 exists as bicarbonate ions (HCO_2^-). This reaction sequence also releases hydrogen ions (H^+), leading to **ocean acidification**. Although CO_2 is only moderately soluble in water, the formation of bicarbonate and carbonate ions allows the oceans and rivers to absorb large quantities of CO_2.

However, in the modern era, the rate of CO_2 emissions and the rate of absorption by surface seawater have exceeded the rate at which bicarbonate can be transported to deeper ocean layers. As a result, acidification of surface waters, where most marine life resides, has become an increasingly severe environmental concern. This illustrates that climate change is not only a terrestrial issue—it directly affects marine ecosystems as well.

On land, the behavior of CO_2 sinks has also changed significantly since the pre-industrial era. Prior to 1850, terrestrial plants absorbed approximately 120–130 $GtCO_2$ per year through photosynthesis, but nearly the same amount was re-emitted into the atmosphere through plant respiration, decomposition, and microbial processes. Thus, the net CO_2 uptake was only about 2–3 Gt per year, representing a relatively balanced biological carbon cycle.

Today, however, the net uptake of CO_2 by vegetation is estimated to have increased to about 11 Gt per year, due in part to CO_2 fertilization—the phenomenon whereby elevated atmospheric CO_2 enhances plant growth. This is a significant contribution when compared to current global CO_2 emissions of around 37 Gt per year. As such, afforestation and ecosystem restoration are widely recognized as effective measures for CO_2 mitigation.

The biological carbon cycle consists of photosynthetic fixation of CO_2 into organic matter, followed by its eventual return to the atmosphere via respiration and decay. It is estimated that about 50% of the CO_2 released during this cycle originates from plant respiration itself. This dynamic cycle functioned with relative stability for millennia, up until the nineteenth century.

However, the recent rise in atmospheric CO_2 due to anthropogenic activities has disturbed this equilibrium. Plants are now growing more rapidly in many regions, altering natural ecosystem processes. These shifts are part of a broader pattern of climate system disruption caused by global warming.

Temperature increases are already producing a range of environmental consequences, including more frequent extreme weather events, accelerated glacial and permafrost melting, sea-level rise, and widespread ecosystem and biodiversity loss. These impacts are occurring across all regions of the Earth and are affecting key human systems such as agriculture, water resources, and public health.

If greenhouse gas emissions continue at current rates, global average temperatures are projected to rise by between **1.5 °C and more than 4 °C by 2100**, depending on the emission scenario. In response to this threat, the **Paris Agreement** established an international objective to limit global temperature rise to **well below 2 °C**, with an aspirational target of **1.5 °C**, relative to pre-industrial levels.

As shown in Fig. 1.7, the main drivers of contemporary global warming are anthropogenic greenhouse gases, particularly carbon dioxide (CO_2), **methane (CH_4)**, and **nitrous oxide (N_2O)** [10]. These gases are primarily emitted through fossil fuel combustion, agricultural activities, and land-use changes, including deforestation. Their accumulation in the atmosphere enhances the greenhouse effect, exacerbating the climate crisis.

1.2 Concerned Situations: Major Impacts of Global Warming

In this section, we summarize the major environmental and societal consequences associated with global warming. The key areas of concern include:

i. **Sea-level rise** due to the melting of glaciers and polar ice sheets

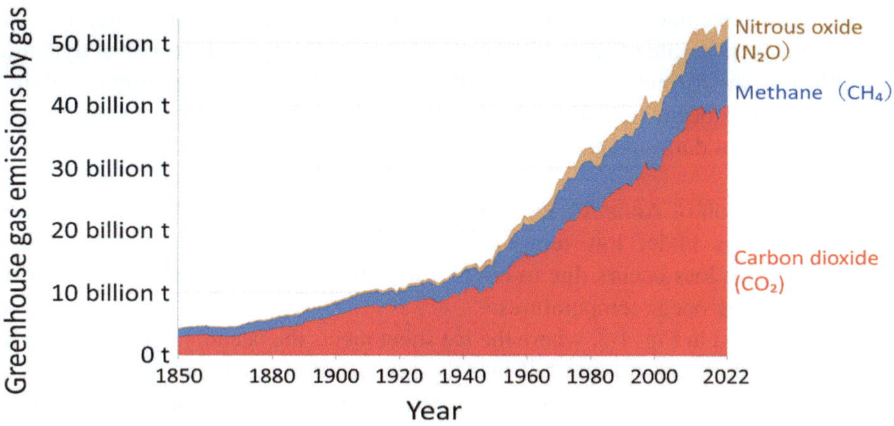

Fig. 1.7 Greenhouse gas emission [10]. (Reprinted from https://ourworldindata.org/grapher/ghg-emissions-by-gas under the Creative Commons Attribution 4.0 International License)

ii. **Increased frequency and severity of extreme weather events**—such as hurricanes, typhoons, storms, heatwaves, and droughts
iii. **Disruption of ecosystems and loss of biodiversity**
iv. **Negative impacts on human health**

Sea-Level Rise

Sea-level rise is one of the most significant and far-reaching consequences of global warming, and it represents a critical global challenge [11–13].

In the Arctic region, most ice exists in the form of **sea ice**, which floats on the ocean surface. Because this ice is already displacing seawater, its melting does not contribute significantly to sea-level rise. However, the situation is markedly different in **Antarctica**, where massive **ice sheets** rest on solid land. These land-based ice sheets are composed of thick layers of compacted snow and ice, with an average thickness of approximately **1900 meters** and maximum thickness reaching **4800 meters** in some areas.

The Antarctic continent spans an area of approximately 14 million square kilometers, representing about 10% of Earth's land surface. If the entire Antarctic ice sheet were to melt, it is estimated that global sea levels would rise by approximately 58.3 meters. The Antarctic ice sheet alone accounts for roughly 90% of the world's glacier mass and over 60% of Earth's freshwater reserves.

These ice sheets also serve an important function in regulating seawater salinity by delivering freshwater into the ocean. This freshwater input influences ocean circulation and tidal dynamics, making the Antarctic cryosphere a crucial component of the global climate system.

Between 1992 and 2016, satellite observations indicate that Antarctica lost ice mass at an average rate of approximately 100 gigatons per year. In a state of dynamic equilibrium, the formation of new ice through snowfall and the melting or calving of ice into the ocean would be balanced, resulting in no net sea-level change. However, recent data suggest that ice loss is accelerating, indicating a disruption in this balance.

The formation of Antarctic ice sheets begins with snowfall, which accumulates and compresses under low temperatures and high pressure to form solid ice. Conversely, ice loss occurs due to both rising atmospheric temperatures and, critically, increasing ocean temperatures.

As illustrated in Fig. 1.8, where the ice sheet meets the ocean, large fragments of ice periodically break off and fall into the sea—forming icebergs. This process is known as calving. Warmer seawater accelerates the rate of calving, and if melting

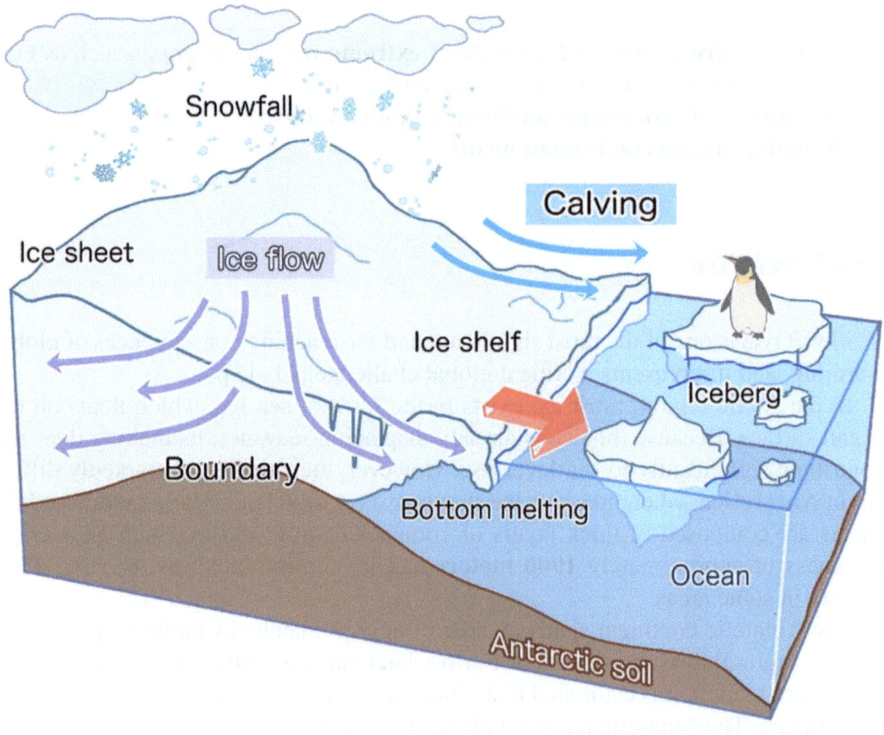

Fig. 1.8 The process of ice sheet formation and melting into the ocean. If the rates of ice accumulation and melting are balanced, the total volume of ocean water remains unchanged. The term *calving* refers to the process by which chunks of ice break off from the edge of an ice sheet or glacier and enter the ocean, contributing freshwater input [14]. Ice shelves often extend into the ocean, and their protruding sections are shaped by surface melting from warm seawater and mechanical erosion by waves. In recent years, the melting rate has increased significantly due to rising ocean temperatures

outpaces formation, the overall volume of ice steadily declines. This imbalance leads to an ongoing rise in global sea levels.

Historically, the **average rate of sea-level rise** over the past 3000 years is estimated to have been relatively modest, at approximately **0.1–0.2 mm per year**. However, according to the **IPCC Sixth Assessment Report (AR6)**, this rate has increased significantly in recent times:

- From **1901 to 1971**, sea levels rose at an average rate of **1.3 mm/year**
- From **1971 to 2006**, the rate increased to **1.9 mm/year**
- From **2006 to 2018**, the rate accelerated further to **3.7 mm/year**

This accelerating trend clearly correlates with the observed rise in global temperatures and the increase in greenhouse gas emissions. If global warming continues unchecked, **sea levels are projected to rise by 20–50 centimeters** over the next 50 years.

Even a seemingly modest rise of 30 cm can have catastrophic consequences, such as frequent coastal flooding, storm surges, and levee failures. In practical terms, this would render existing coastal infrastructure inadequate, requiring a complete redesign of disaster prevention and flood management systems.

Several major coastal cities are already facing high risks of submersion due to sea-level rise and land subsidence. These include:

- Jakarta
- Miami
- Ho Chi Minh City
- Mumbai
- Guangzhou
- Shanghai
- New York
- Bangkok
- Manila
- Tokyo

Projections indicate that, without substantial mitigation of greenhouse gas emissions, global mean sea level could rise by approximately 80 centimeters by the end of the twenty-first century (2100). Furthermore, under worst-case scenarios extending beyond this century, sea-level rise could exceed 15 meters by the year 2500, largely due to the potential collapse of major ice sheets in Antarctica and Greenland.

These data underscore the severity of global warming and its long-term consequences. To avoid irreversible damage to ecosystems, infrastructure, and human livelihoods, it is imperative to drastically reduce greenhouse gas emissions starting immediately.

Extreme Weather Events and Their Link to Global Warming

Climate phenomena such as **hurricanes**, **typhoons**, **storms**, **heatwaves**, and **droughts** are increasingly influenced by global warming, primarily due to alterations in the Earth's energy balance resulting from the enhanced greenhouse effect. Among these, the intensification of tropical cyclones—including hurricanes and typhoons—is one of the most prominent examples. The primary driver of cyclone intensification is the rise in sea surface temperatures (SSTs). Warmer SSTs increase the rate of evaporation, injecting more moisture into the atmosphere. As this moisture-laden air rises, it initiates deep convection, during which latent heat is absorbed through the evaporation of seawater. The water vapor subsequently condenses at higher altitudes, releasing latent heat and warming the surrounding air. This release of latent heat enhances vertical air motion, generating stronger updrafts that draw in additional air from the surrounding environment. This positive feedback mechanism constitutes the energy engine that fuels hurricanes and typhoons. As atmospheric temperature gradients steepen, vertical convection intensifies, promoting more efficient transfer of thermal energy into the upper troposphere. Although adiabatic cooling occurs during ascent, the release of latent heat partially offsets this effect, allowing the storm system to sustain and intensify.

Additionally, elevated SSTs contribute to lower central pressures within cyclonic systems, enhancing the inflow of air from surrounding high-pressure regions and increasing wind speeds. Stronger winds amplify the destructive potential of storms. The overall energy available to these systems increases with greater latent heat flux—the transfer of energy from the ocean to the atmosphere via evaporation—which further strengthens storm intensity [15].

In parallel with intensified storms, the frequency and severity of droughts and heatwaves are also increasing [16, 17]. Rising global temperatures, driven by greenhouse gases such as CO_2 and CH_4, elevate atmospheric water vapor content and enhance evaporation from the Earth's surface. In many arid and semi-arid regions, this leads to evaporation exceeding precipitation, resulting in the drying of soils and vegetation.

Global warming also alters precipitation patterns, making rainfall more erratic. While some regions experience increased precipitation, others suffer from prolonged dry periods. Moreover, extreme rainfall events are becoming more intense but shorter in duration, often followed by extended dry spells, exacerbating drought conditions.

The Intertropical Convergence Zone (ITCZ)—a key driver of equatorial rainfall—is also influenced by climate change. Shifts in the ITCZ can significantly alter regional precipitation distributions. For instance, a northward displacement may cause droughts in southern tropical regions, while a southward shift may lead to drying in northern tropical areas. Additionally, the persistence and intensification of high-pressure systems over specific regions can inhibit cloud formation and precipitation, further reinforcing drought conditions. High surface temperatures accelerate soil moisture evaporation, reduce vegetation cover, and decrease agricultural yields.

Ecosystem Disruption and Biodiversity Loss

These climate changes also pose serious threats to ecosystems and biodiversity [18]. Many species are highly sensitive to temperature and precipitation changes, especially in vulnerable ecosystems such as coral reefs, Arctic tundra, and tropical forests. Rapid climate shifts can lead to habitat loss, population declines, and species extinctions. The loss of biodiversity undermines ecosystem services—such as food security, freshwater supply, pollination, and climate regulation—which are vital for human well-being and economic stability.

Impacts on Human Health and Society

Global warming also has direct and indirect consequences for human health. Elevated temperatures are linked to increased incidence of **heatstroke**, **cardiovascular events**, and **respiratory illnesses**. Warmer climates expand the habitable range of **disease vectors** such as mosquitoes and ticks, raising the risk of infectious diseases including **malaria**, **dengue fever**, and **Lyme disease**.

Water scarcity and food insecurity—exacerbated by droughts and floods—contribute to **malnutrition**, **waterborne diseases**, and public health crises. Moreover, climate-related disasters and ecosystem degradation disrupt critical sectors such as agriculture, fisheries, and tourism. These disruptions result in infrastructure damage, increased healthcare demand, and substantial economic losses, placing significant burdens on both developed and developing societies.

1.3 Scientific Evidence for Global Warming Due to Anthropogenic CO_2 Emissions

It is now established with a high degree of scientific certainty that the rise in global temperature observed since the 1980s is primarily attributable to the large-scale emission of the greenhouse gas carbon dioxide (CO_2) by human activities. A comprehensive study published in 2021 examined whether the scientific literature supports the view that recent climate change is anthropogenic, or whether there exists a significant body of skeptical peer-reviewed research [19]. The paper, titled "Greater Than 99% Consensus on Human-Caused Climate Change in the Peer-Reviewed Scientific Literature," presents a statistical analysis of the scientific consensus, as illustrated in Fig. 1.9. Between 2012 and 2020, a total of 88,125 peer-reviewed articles related to recent climate change were published. Lynas et al. randomly selected a representative sample of 3000 papers from this dataset and conducted a keyword search across the full dataset to identify papers expressing skepticism regarding human-induced climate change. Remarkably, only four papers in the

ENVIRONMENTAL RESEARCH
LETTERS

LETTER • OPEN ACCESS

Greater than 99% consensus on human caused climate change in the peer-reviewed scientific Literature

Mark Lynas et al 2021 Environ. Res. Lett. 16 114006

Fig. 1.9 The title of a paper published in Environmental Research Letters (2021) [19]

3000-sample subset were found to express skeptical views. This corresponds to just 0.1% of the literature, indicating that approximately 99.9% of the peer-reviewed scientific research supports the conclusion that recent global warming is caused by human activity. In summary, there is an overwhelming scientific consensus: the current trajectory of climate change is anthropogenic in origin.

Scientific data linking global warming to the increase in atmospheric carbon dioxide (CO_2) concentrations since the Industrial Revolution are robust and derived from multiple independent lines of research and observation. The main categories of scientific evidence for anthropogenic global warming are outlined below:

i. **Direct Observations of CO_2 Concentrations**

- **Long-Term Atmospheric Measurements**: Continuous measurements at the Mauna Loa Observatory in Hawaii, initiated in 1958, have documented a steady rise in atmospheric CO_2 levels. This record, known as the *Keeling Curve*, shows that CO_2 concentrations have increased from approximately 280 ppm in the pre-industrial era to over 420 ppm today.
- **Ice Core Data**: Ice cores extracted from Antarctica and Greenland provide atmospheric records spanning the past 800,000 years. These cores reveal that current CO_2 levels far exceed the natural fluctuations observed in previous glacial and interglacial cycles.

ii. **Physics of the Greenhouse Effect**

- **Infrared Absorption Properties of Greenhouse Gases**: Laboratory experiments have conclusively demonstrated that CO_2 and other greenhouse gases absorb and emit infrared radiation. As atmospheric CO_2 concentrations rise, more infrared radiation emitted from the Earth's surface is absorbed and re-radiated back, causing surface and lower atmospheric warming. This phenomenon is well explained by the principles of radiative transfer and molecular spectroscopy.

iii. **Changes in Earth's Energy Balance**

- **Satellite Measurements**: Satellite observations of Earth's radiation budget (net radiation balance) have revealed a decrease in the amount of outgoing longwave radiation escaping into space. This reduction corresponds with the rise in atmospheric CO_2 concentrations, suggesting that more thermal energy is being retained within the Earth system due to the enhanced greenhouse effect. According to satellite data from the CERES (Clouds and the Earth's Radiant Energy System) mission, the Earth is currently experiencing a persistent energy imbalance of approximately 0.6–0.8 watts per square meter (W/m^2). This excess energy, caused primarily by the accumulation of greenhouse gases, is absorbed by the atmosphere, land, and particularly the oceans—more than 90% of which is stored in the upper ocean layers. This imbalance is a key indicator of ongoing global warming, even in the presence of short-term natural variability.

iv. **Empirical Evidence of Global Warming**

- **Rising Global Temperatures**: Instrumental records show a clear upward trend in global average surface temperatures over recent decades, with the rate of warming accelerating particularly since the mid-twentieth century. This temperature increase closely correlates with rising atmospheric CO_2 levels.
- **Climate Model Simulations**: Multiple climate models have consistently demonstrated that the observed warming cannot be explained by natural variability alone. These models identify rising CO_2 concentrations as a primary driver of recent global warming, reinforcing the causal relationship.

These lines of evidence strongly support the conclusion that CO_2-induced enhancement of the greenhouse effect is a major contributor to global warming. The increase in atmospheric CO_2 is a directly measured and undisputed fact, even when accounting for seasonal fluctuations. While it is true that some uncertainties remain regarding the precise attribution of short-term global temperature variations, the long-term warming trend is well understood. What is clear is that without further increases in CO_2, the associated warming effect can be halted. However, once CO_2 levels rise, the resulting climatic changes are not easily reversible. This underscores a fundamental reality: the accumulation of CO_2 in the atmosphere is a one-way process within human timescales. Recognizing this irreversibility is essential for understanding the urgency of mitigation strategies in addressing climate change.

1.4 Misunderstanding of the Mechanism of Global Warming

One major challenge in addressing climate change is the widespread misunderstanding of the physical mechanism underlying global warming. A clearer understanding of this mechanism helps to appreciate the seriousness of the issue.

In the Northern Hemisphere, it is commonly observed that cloudy nights in winter tend to be warmer than clear ones. This is due to the *radiative cooling effect*: when skies are clear, heat from the Earth's surface escapes efficiently into space; when clouds are present, water vapor (H_2O) absorbs and re-emits infrared radiation, trapping heat and keeping the surface warmer. This is a manifestation of the greenhouse effect. The term "greenhouse" is a helpful analogy, though somewhat simplified. As illustrated in Fig. 1.10, clouds act like a greenhouse with holes—some infrared radiation escapes through gaps. In contrast, carbon dioxide (CO_2) functions more like a sealed greenhouse, because it is evenly distributed throughout the atmosphere—from tens of meters above the surface up to the tropopause (~10 km)—effectively trapping heat over a broad vertical range. Unlike clouds, which vary greatly in time and location and primarily affect local temperatures through short-term feedback mechanisms, CO_2 is a long-lived greenhouse gas that is uniformly mixed throughout the troposphere and persists in the atmosphere for decades to centuries. Its radiative effect is continuous and global in scope. While clouds can both warm and cool the Earth depending on their type, altitude, and time of day, the effect of CO_2 is unidirectional: it traps outgoing infrared radiation and contributes consistently to long-term global warming. This fundamental distinction explains why the impact of CO_2 is persistent and globally uniform. This image helps conceptualize the persistent and widespread impact of CO_2. Simply put: the more CO_2 there is in the atmosphere, the more heat is retained near the surface, leading naturally to a rise in global temperatures. Let us now explain the mechanism of global warming more concretely.

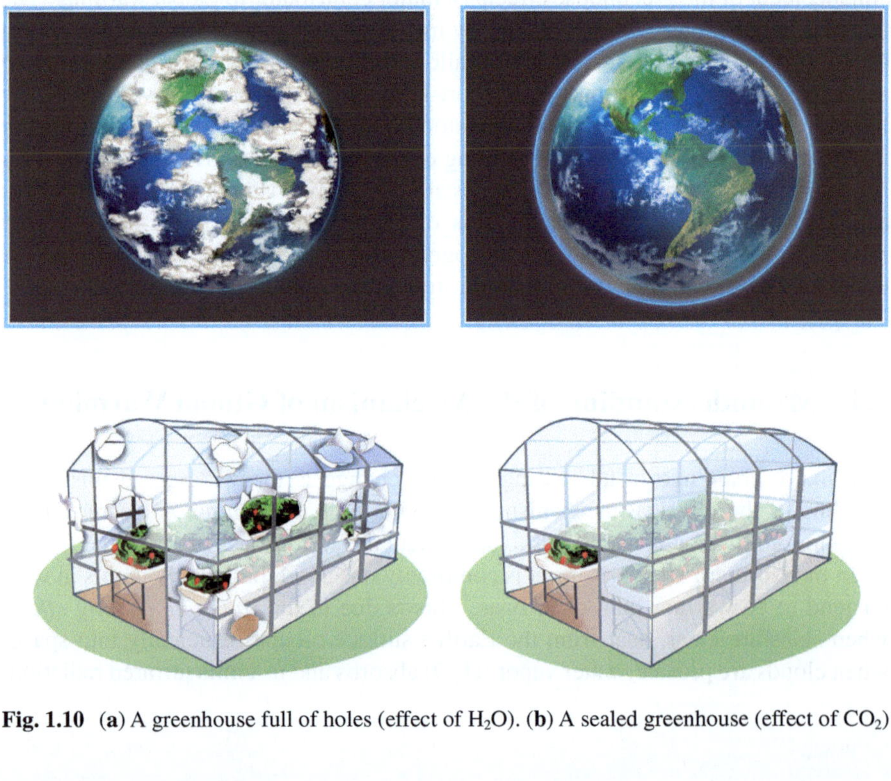

Fig. 1.10 (**a**) A greenhouse full of holes (effect of H_2O). (**b**) A sealed greenhouse (effect of CO_2)

The Role of Radiation in the Greenhouse Effect

To understand Earth's greenhouse effect, it is essential to distinguish between two types of radiation:

i. **Shortwave radiation** from the Sun, primarily in the form of ultraviolet, visible, and near-infrared light (wavelengths ~0.1 to 5 μm), which penetrates the atmosphere and warms the Earth's surface.
ii. **Longwave (infrared) radiation** emitted by the Earth's surface (wavelengths ~5 to 100 μm), which carries heat back toward space.

The difference in wavelengths arises from the temperatures of the emitting bodies. The Sun, at approximately 6000 °C, emits mostly shortwave radiation. In contrast, the Earth, with an average surface temperature of about 14 °C, emits longwave infrared radiation as blackbody radiation. As shown in Fig. 1.11, the solar spectrum includes ultraviolet, visible, and near-infrared light. A portion of this radiation—particularly near-infrared (0.8–3 μm)—is absorbed by water vapor and CO_2, especially near the surface. Oceans, which cover about 70% of the Earth's surface, are particularly effective at absorbing this energy. Ultraviolet radiation (wavelengths <0.4 μm) is largely absorbed in the upper atmosphere, particularly by the ozone layer at around 11 km altitude. The surface, after being warmed by incoming solar radiation, emits infrared radiation upward. This outgoing radiation is then partially absorbed and re-emitted by greenhouse gases such as water vapor, CO_2, CH_4, CFCs, N_2O, and O_2. These gases trap heat within the atmosphere by re-radiating infrared energy both upward and downward, effectively warming the troposphere and maintaining Earth's surface temperature. Without the greenhouse effect, the Earth's average surface temperature would be approximately −18 °C, making it inhospitable for life as we know it.

Infrared Absorption by CO_2 Molecules

Let us briefly examine how CO_2 absorbs infrared radiation at the molecular level. CO_2 does not absorb radiation in the visible light range (approximately 380–700 nm). However, it strongly absorbs infrared radiation at specific wavelengths: around 2.7 μm, 4.3 μm, and 15 μm. These absorption bands correspond to distinct vibrational modes of the CO_2 molecule:

- **15 μm**: bending vibration (asymmetric deformation of the molecule)
- **4.3 μm**: asymmetric stretching vibration
- **2.7 μm**: combination of symmetric and asymmetric stretching modes

Of these, the **bending vibration at 15 μm** is especially relevant for climate change because this wavelength region is not significantly absorbed by water vapor, making CO_2 the dominant absorber at this band. As illustrated in Figs. 1.12

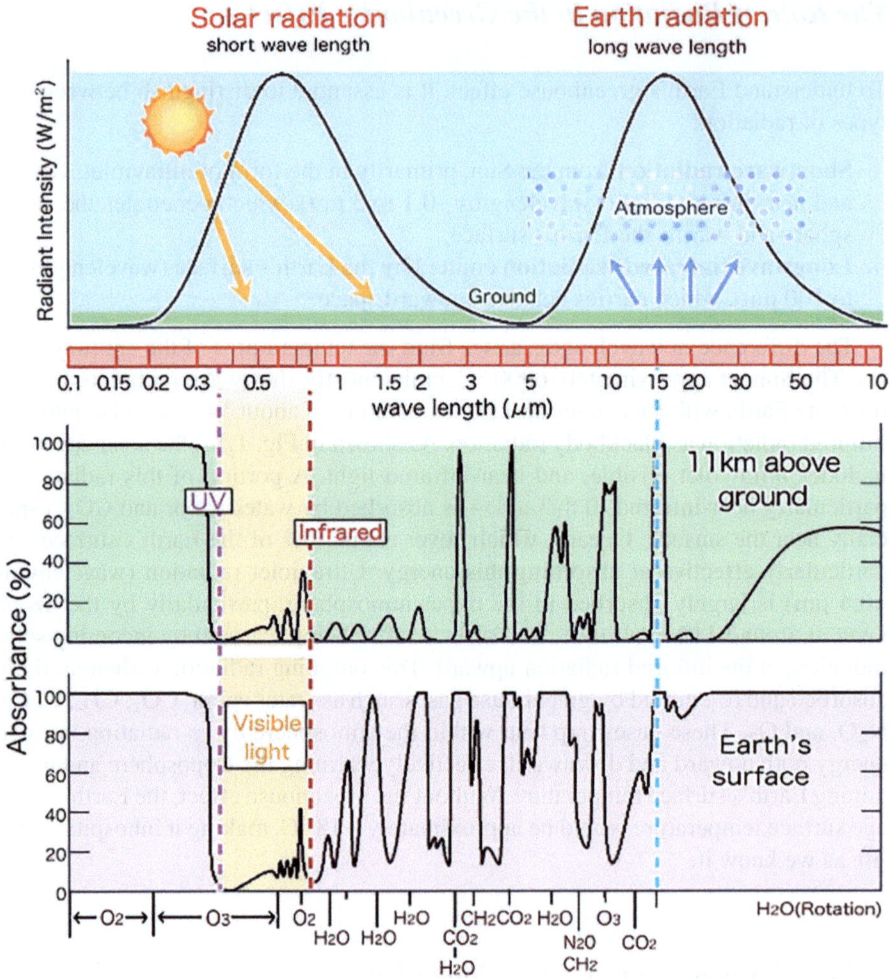

Fig. 1.11 Differences in wavelength ranges between solar radiation and earth radiation

(infrared absorption spectrum) and 1.13 (molecular motion diagram), the energy of vibrational motion lies in the infrared region and is largely responsible for the warming effect of infrared radiation—similar to how an infrared heater works. When CO_2 absorbs infrared radiation, the energy is stored in the molecule's vibrational motion. The energized CO_2 molecules then collide with other atmospheric gases such as N_2 and O_2, transferring heat through molecular collisions. This process increases the thermal energy (temperature) of the surrounding air. This mechanism of infrared absorption and re-emission is what makes CO_2 such an effective greenhouse gas. As with the earlier example of warmth retained in a plastic greenhouse (e.g., a vinyl house), the transparent material lets in shortwave sunlight but traps longwave infrared radiation—an analogy for how CO_2 behaves in the atmosphere (Fig. 1.13).

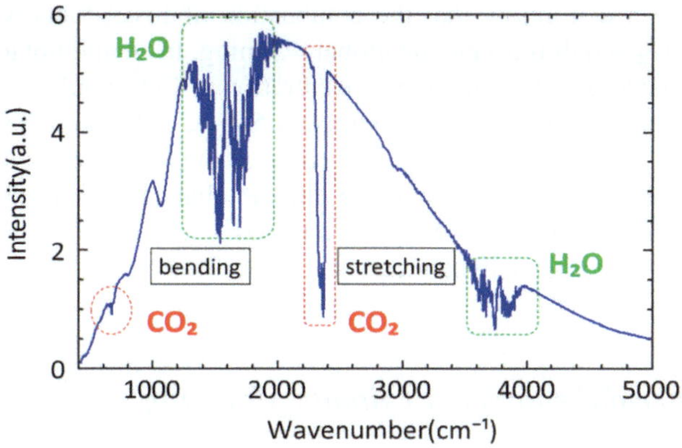

Fig. 1.12 Infrared spectrum of CO_2 and H_2O

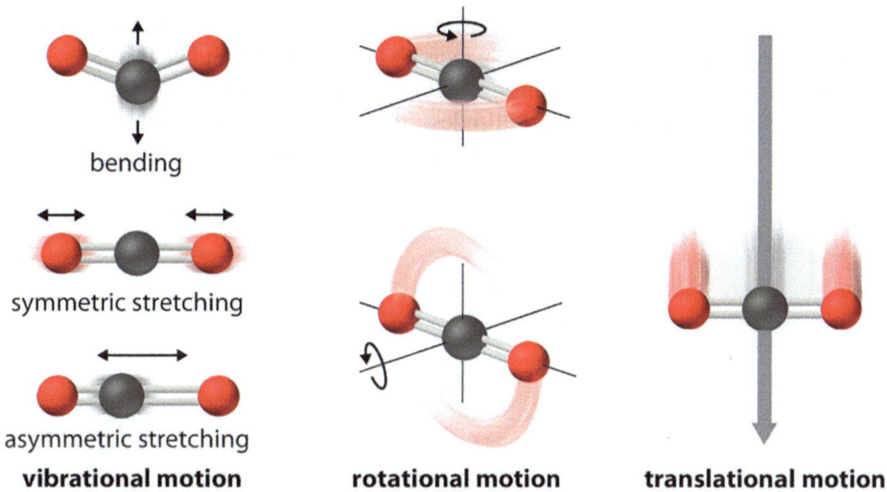

Fig. 1.13 Modes of motion of the CO_2 molecule, i.e., vibrational, rotational, and translational modes

Early Scientific Recognition of CO_2-Induced Warming

Swedish chemist Svante Arrhenius, best known for his work on chemical reaction rates, proposed as early as 1896 that an increase in atmospheric carbon dioxide (CO_2) could lead to a rise in Earth's surface temperature [20]. He recognized CO_2 as a greenhouse gas, capable of absorbing infrared radiation and thereby contributing to the warming of the Earth's surface. Arrhenius's work was among the first scientific studies to explore the relationship between atmospheric CO_2 concentration and global temperature, laying the groundwork for modern climate science.

Remarkably, he also foresaw that the combustion of fossil fuels would increase atmospheric CO_2 and thus accelerate global warming. In an attempt to quantify this effect, he estimated that doubling the concentration of CO_2 would result in a global average temperature increase of approximately 5–6 °C. Although this estimate was relatively rough by today's standards, it introduced the fundamental idea that atmospheric CO_2 levels are directly linked to climate change. His theoretical insights deserve recognition as foundational contributions to our current understanding of global warming and climate dynamics.

Manabe's Breakthroughs in Climate Modeling

As mentioned above, the mechanism of global warming can be explained by the repeated absorption and re-emission of radiation, starting with incoming solar radiation, as illustrated in Fig. 1.14. Solar radiation warms the Earth's surface, which then emits infrared radiation. This outgoing infrared radiation is partially absorbed and re-emitted by greenhouse gases such as CO_2, CH_4, H_2O, and N_2O.

As the atmospheric concentration of CO_2 increases, more infrared radiation is absorbed and re-emitted at higher altitudes. Since these higher altitudes are colder, the amount of energy radiated back into space decreases, causing more heat to be

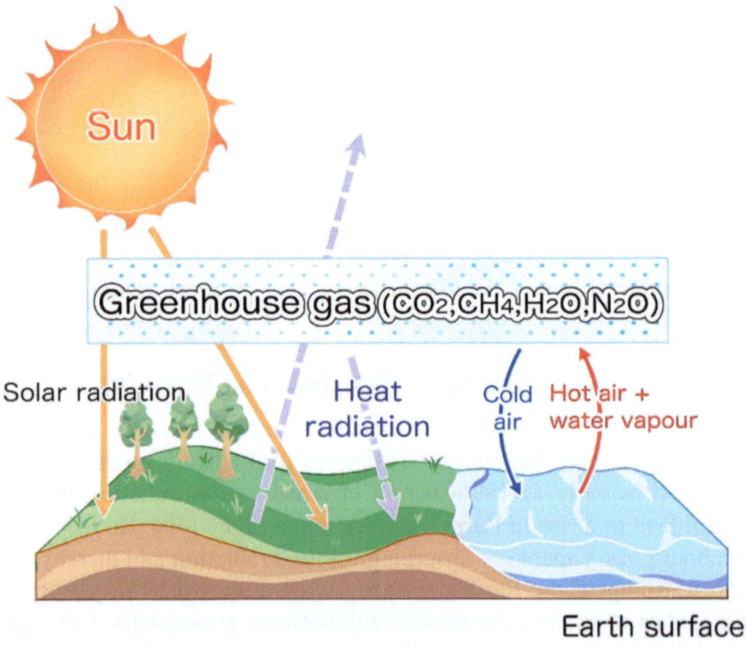

Fig. 1.14 Mechanism of global warming by greenhouse gases

retained within the Earth's climate system. This process leads to an overall increase in the Earth's surface temperature.

In 1964, Syukuro Manabe, co-recipient of the 2021 Nobel Prize in Physics, introduced the one-dimensional radiative-convective equilibrium model to simulate the greenhouse effect [21]. This model treats the atmosphere as a vertical column extending from the Earth's surface to the upper atmosphere, without accounting for horizontal variation. It captures the steady-state energy balance between solar radiation absorbed at the surface and infrared radiation emitted from the Earth back into space. As discussed earlier, greenhouse gases trap outgoing infrared radiation, and the retained heat is redistributed vertically through atmospheric convection. This energy exchange is primarily facilitated by collisions between air molecules. In a landmark 1967 paper, Manabe demonstrated that doubling atmospheric CO_2 would lead to an increase in global average surface temperature of about 2.3 °C, relative to a baseline climate with pre-industrial CO_2 levels. This was one of the first quantitative estimates of the climate sensitivity to CO_2 and it marked a turning point in climate modeling. Later, Manabe developed a more advanced coupled atmosphere-ocean model to investigate the role of the ocean in climate dynamics [22]. Because water has a high heat capacity, the ocean heats and cools slowly, acting as a thermal buffer. It is now understood that the ocean absorbs more than 90% of the excess energy retained by the Earth system due to greenhouse gases. Manabe's model marked the beginning of physics-based climate modeling and served as a prototype for the comprehensive models later adopted by the IPCC. These models—now run on modern supercomputers—form the backbone of today's climate simulations. They are considered highly reliable tools for projecting future climate scenarios and assessing the impacts of anthropogenic greenhouse gas emissions.

As mentioned above, there is a scientifically established relationship between the concentration of CO_2 in the atmosphere and temperature, and this relationship is one of the basic elements in the earth's climate system. This basic scientific knowledge is essential for understanding and addressing current global warming and its impacts.

References

1. Climate Change—Our World in Data, Met Office Hadley Centre—Hadley Centre's Sea Surface Temperature (HadSST) (2025)—with major processing by Our World in Data. https://ourworldindata.org/climate-change. 7 June 2025
2. Wikipedia contributors. Global average temperature—NASA-GISS, HadCrut, NOAA, Japan, Berkeley Earth, Wikimedia Commons. Licensed under CC BY-SA 4.0 International. https://en.wikipedia.org/wiki/File:20200324_Global_average_temperature_-_NASA-GISS_HadCrut_NOAA_Japan_BerkeleyE.svg. Accessed 7 June 2025
3. IPCC Fourth Assessment Report: climate Change 2007, Working Group I: The Physical Science Basis

4. Our World in Data. Emissions by fuel. Hannah Ritchie and Max Roser (2024). Published at https://ourworldindata.org/emissions-by-fuel. Licensed under CC BY 4.0.CO_2 emissions by fuel—Our World in Data, Accessed 7 June 2025
5. Luthi, D., et al. (2008) Etheridge, D.M., et al. 2010; Vostok ice core data/J.R. Petit et al.; NOAA Mauna Loa CO_2 record. Trends in CO_2—NOAA Global Monitoring Laboratory. Accessed 1 Sept 2025
6. NASA Global Climate Change: Vital Signs of the Planet, Graphic: The relentless rise of carbon dioxide—NASA Science. Accessed 1 September 2025
7. IEA report 2023: CO_2 Emissions in 2023. https://www.iea.org/reports/CO_2-emissions-in-2023. License: CC BY 4.0
8. Friedlingstein P et al (2023) Earth Syst Sci Data 14:5301–5369
9. Earth Syst Sci Data. 14, 4811–4900, 2022 (p. 4834, 4835)
10. Climate Change—Our World in Data, Greenhouse gas emissions by gas, World, 1850 to 2023, Jones et al. (2024)—with major processing by Our World in Data, CC BY https://ourworldindata.org/grapher/ghg-emissions-by-gas. Accessed 30 Dec 2024
11. Sugiyama S (2021) What is happening to the ice in Antarctica? The science of climate change and ice sheets. Chuokoron-Shinsha, Tokyo
12. Alley RB (2002) The two-mile time machine: ice cores, abrupt climate change, and our future. Princeton University Press, Princeton
13. Jouzel J, Lorius C, Raynaud D (2016) The white planet: the evolution and future of our frozen world. Princeton University Press, Princeton
14. Mosbeux C, Wagner TJW, Becker MK, Fricker HA (2020) Viscous and elastic buoyancy stresses as drivers of ice-shelf calving. J Glaciol 66:643–657. https://doi.org/10.1017/jog.2020.35
15. Knutson TR, McBride JL, Chan J, Emanuel K, Holland G, Landsea C, Held I, Kossin JP, Srivastava AK, Sugi M (2010) Tropical cyclones and climate change. Nat Geosci 3:157–163
16. Dai A (2013) Increasing drought under global warming in observations and models. Nat Climate Change 3:52–58
17. Trenberth KE, Dai A, Schrier G, Jones PD, Barichivich J, Briffa KR, Sheffield J (2014) Global warming and changes in drought. Nat Climate Change 4:17–22
18. IPCC Special Report (2019) Climate change and land. IPCC. https://www.ipcc.ch/srccl/. Accessed 30 Dec 2024
19. Lynas M, Houlton BZ, Perry S (2021) Greater than 99% consensus on human caused climate change in the peer-reviewed scientific literature. Environ Res Lett 16:114005. https://doi.org/10.1088/1748-9326/ac2966
20. Arrhenius S (1896) On the influence of carbonic acid in the air upon the temperature of the ground. Philos Magazine J Sci Ser 5(41):237–276
21. Manabe S, Strickler RF (1964) Thermal equilibrium of the atmosphere with a convective adjustment. J Atmos Sci 21:361–385
22. Manabe S, Bryan K (1969) Climate calculation with a combined ocean-atmosphere model. J Atmos Sci 26:786–789

Chapter 2
Energy Issues

Abstract This chapter examines the central role of fossil fuels in climate change and the urgent need for a fundamental shift in energy systems. It begins by distinguishing between primary and secondary energy, highlighting that fossil fuels still account for over 80% of global primary energy consumption. This chapter stresses that CO_2 emissions scale almost linearly with fossil energy use. It then discusses the promise and limitations of renewable energy, focusing on solar and wind, and identifies the need for energy carriers to address intermittency and transport issues. While batteries offer some solutions, they face significant limitations in cost, scale, and resource availability. This chapter introduces various chemical energy carriers, such as hydrogen, ammonia, methane, and especially methanol, and evaluates their suitability using quantitative criteria like energy density, infrastructure compatibility, and environmental impact. This chapter emphasizes that the selection of future energy carriers must be grounded in data and systems-level thinking.

Keywords Combustion of fossil fuels · Renewable energy · Energy carriers · Hydrogen · Methanol · Ammonia · Energy density

2.1 The Link Between Energy Use and Climate Change

A quantitative understanding of the link between energy use and climate change is essential. It forms the basis for informed discussions and policy decisions about how humanity should respond to future climate risks. To begin, let us clarify the types of energy we use. Energy can be broadly classified into primary and secondary energy:

- Primary energy refers to energy directly extracted from natural sources—such as oil, natural gas, coal, hydropower, nuclear power, and other renewables.
- Secondary energy includes electricity, city gas, gasoline, and kerosene, which are produced by converting and processing primary energy for end use.

J. Nakamura, M. Ishida, *Blueprint for a Methanol Society*,
https://doi.org/10.1007/978-981-95-3465-4_2

For instance, electricity generated through the combustion of coal or natural gas is a typical form of secondary energy. According to data from the Energy Institute, the global total of primary energy consumption was 568 EJ (exa = 10^{18}) in 2020, 597 EJ in 2021, 607 EJ in 2022, and 620 EJ in 2023 [1]. In 2023, while primary energy totaled 620 EJ, the amount of usable secondary energy was only 445 EJ [2]. The difference—175 EJ—represents energy losses and conversion costs. This means that approximately 71.8% of primary energy is converted into usable secondary energy, while the remaining 28.2% is lost during energy conversion, transportation, and distribution. These losses include heat released during refining processes (e.g., producing gasoline from crude oil) and inefficiencies in power generation and infrastructure operation. Costs also include technical and financial expenditures for building and maintaining power plants, refineries, and distribution networks.

Fossil Fuel Use and CO_2 Emissions

Energy use is the primary cause of anthropogenic climate change. The total energy consumed by society corresponds to primary energy, and thus, CO_2 emissions are directly linked to the consumption of fossil fuels. Figure 2.1 shows the breakdown of primary energy use up to 2022. Prior to 1950, coal was the dominant energy source. Oil use expanded rapidly until 1970, followed by a surge in natural gas consumption. By 2022, coal, oil, and natural gas accounted for 82% of global primary energy—demonstrating our continued reliance on fossil fuels. As also illustrated in Chap. 1 (see Fig. 1.3), annual CO_2 emissions track closely with fossil fuel consumption. This is expected, as fossil fuels consist mainly of carbon and hydrogen. When oxidized to release energy, they produce CO_2 and H_2O according to the generalized combustion reaction:

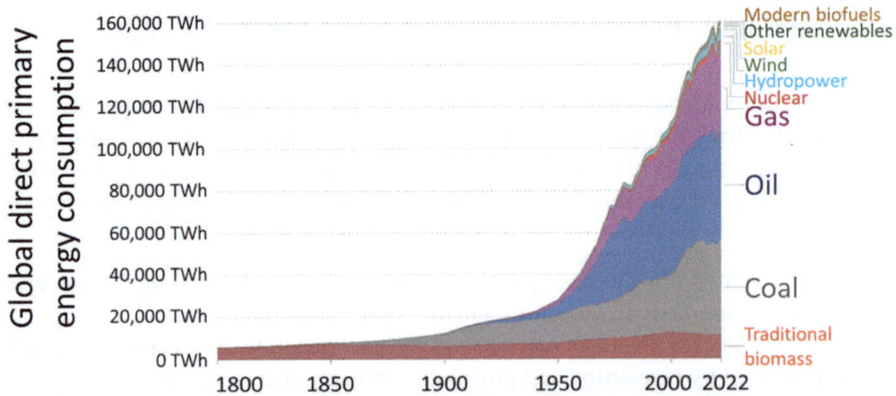

Fig. 2.1 Annual consumption of global primary energy with inset of annual CO_2 emission [3]. Data source; Our World in Data, Reprinted from https://ourworldindata.org/global-energy-200-years, under the Creative Commons Attribution 4.0 International License

$$C_xH_y + \left(x + \frac{y}{4}\right)O_2 \rightarrow xCO_2 + \frac{y}{2}H_2O \tag{2.1}$$

This reaction releases energy—referred to as combustion heat in high school curricula and combustion enthalpy in university-level chemistry. Since fossil fuel consumption and CO_2 emissions are both calculated from fuel mass and composition, there exists an approximately 1:1 relationship between the amount of fossil-derived primary energy used and the resulting CO_2 emissions. Let us estimate global CO_2 emissions from fossil energy in 2022 (600 EJ/year). Based on the hydrogen-to-carbon ratio (y/x), we note:

- Coal: low hydrogen content ($y/x < 1$)
- Oil: hydrocarbons with $y/x \approx 2$
- Natural gas (mainly CH_4): $y/x \approx 4$

Complete combustion of CH_4 releases ~890.4 kJ/mol. In general, 700–900 kJ of energy is released per mole of carbon atoms, each producing 44 g of CO_2. Thus, emitting 35 billion tons of CO_2 corresponds to 560–720 EJ, which aligns well with the 2022 primary energy estimate of 600 EJ. This clearly demonstrates that fossil energy use and CO_2 emissions are directly linked. As of 2024, more than 99% of emitted CO_2 enters the atmosphere because carbon capture, utilization, and storage (CCUS) technologies remain underdeveloped and are not yet deployed at scale. Therefore, unless we dramatically reduce fossil fuel usage, climate change will continue to accelerate.

The Urgent Need for an Energy Transition

Currently, fossil fuels account for 80% of primary energy. Simply reducing electricity consumption or changing individual behavior is not enough to address this challenge. What is required is a large-scale energy revolution—a complete transformation of the global energy system.

There are only three major energy sources: fossil fuels (coal, oil, and natural gas), nuclear power, and renewables. Biomass energy is an important energy source in a sustainable society, but it cannot become the primary energy source. If fossil fuels are to be phased out, humanity must choose between nuclear and renewables.

While nuclear power can provide large-scale energy, concerns over radioactive waste disposal and public acceptance limit its future potential. As a result, the most promising pathway is to transition from fossil fuels to renewables, particularly solar, wind, hydropower, and biomass (hydropower is included here as renewable energy).

According to 2022 data [3], renewables accounted for:

- Solar and wind: 5.3% of primary energy
- Biomass: 6.9% of primary energy

For reference, nuclear power accounted for 3.7%. Biomass is considered carbon neutral, because the CO_2 released upon combustion is offset by the CO_2 absorbed during plant growth. This is why biomass does not appear in the CO_2 emissions diagram (inset of Fig. 2.1).

Figure 2.2 shows data on electricity consumption from renewable energy sources from 1990 to 2021. Until around 2010, hydropower accounted for the majority of renewable electricity generation. Since then, the share of solar and wind power has steadily increased. According to the *Tracking SDG 7* report (2024), total renewable electricity consumption in 2021 reached approximately 25 EJ. It is important to note that this figure represents electricity use—i.e., secondary energy. Under the substitution method used by the IEA and others, renewable electricity is converted to an equivalent amount of primary energy using a reference fossil fuel efficiency, typically assumed to be 40%. Based on this, 25 EJ of electricity would correspond to approximately 62.5 EJ of primary energy. Given that global primary energy consumption in 2021 was about 597 EJ, renewable electricity including hydropower would account for approximately 10.4%. However, other sources such as the Energy Institute report a higher share of around 14%, depending on the inclusion of bioenergy and the accounting methodology used.

Next, let us estimate the potential CO_2 reductions achievable through large-scale expansion of renewable energy. If renewable electricity consumption increases to 125 EJ—five times the 2021 level—this would correspond to 312.5 EJ of primary energy (using the same 40% efficiency assumption). Since global primary energy consumption in 2023 was 620 EJ, this would represent roughly half of the total,

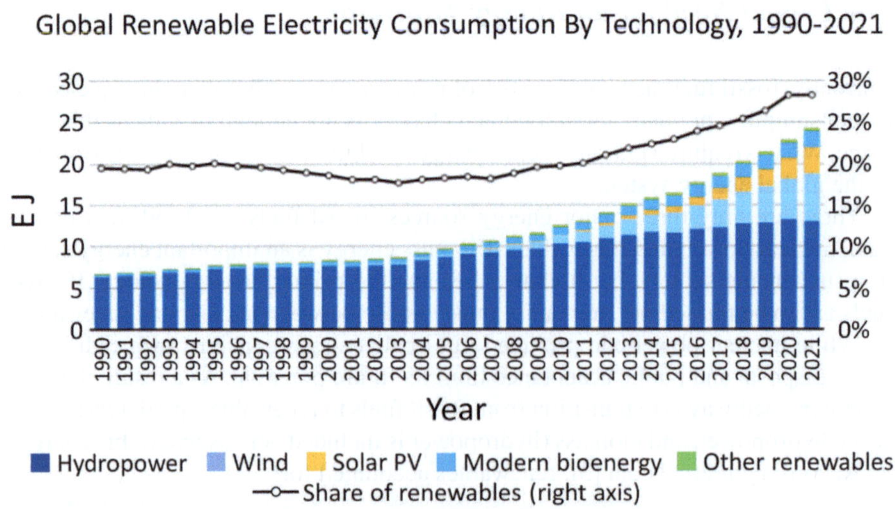

Fig. 2.2 Global renewable electricity consumption by technology and the share of renewable energy in the global electricity consumption. Adapted from [4] with permission by the United Nations Statistics Divisions (UNSD)

suggesting the potential to cut CO_2 emissions by around 50%, assuming fossil energy is replaced.

In principle, deep reductions in CO_2 emissions can be achieved by significantly scaling up renewables, especially in developing countries. Additional reductions are possible if remaining fossil fuel use is combined with CO_2 recycling technologies. The central message of this book is that methanol production using CO_2 and renewable energy represents one of the most effective strategies for achieving substantial emissions reductions.

Energy Conversion Efficiency and the Cost of Electricity

To better appreciate the quality and value of electricity, let us examine the efficiency of converting fossil resources—used as primary energy—into electricity. This will help us understand how much energy is lost as waste heat in the process, and how valuable electricity is as a high-grade energy form.

In developed countries, the conversion efficiency of fossil fuels into electricity varies depending on the technology used:

- Coal-fired power plants typically achieve 30–50% efficiency.
- Natural gas combined cycle (NGCC) plants reach 50–60% efficiency.

This means that for every unit of electricity generated, a comparable amount of energy is lost as heat. In simpler terms, we discard as much energy as we use.

If fossil energy is used directly for heating—such as in boilers or industrial furnaces—the thermal efficiency can be relatively high. However, converting fossil energy into electricity is significantly less efficient due to thermodynamic limitations.

In this sense, electricity is a high-quality and costly form of energy, both in terms of energy investment and infrastructure requirements.

An even more critical issue is that, as previously discussed, CO_2 is emitted during the combustion of fossil fuels in the power generation process. The CO_2 emissions associated with electricity generation are a major contributor to climate change.

To address this, it is essential to transition as much as possible from fossil fuel-based electricity generation to renewable energy sources, which produce electricity without emitting CO_2.

2.2 Challenges in Integrating Renewable Energy

In the previous section, we discussed the essential role of renewable energy in future energy systems. However, the core issue is not simply recognizing the importance of renewables, but rather determining how they should be incorporated and how to

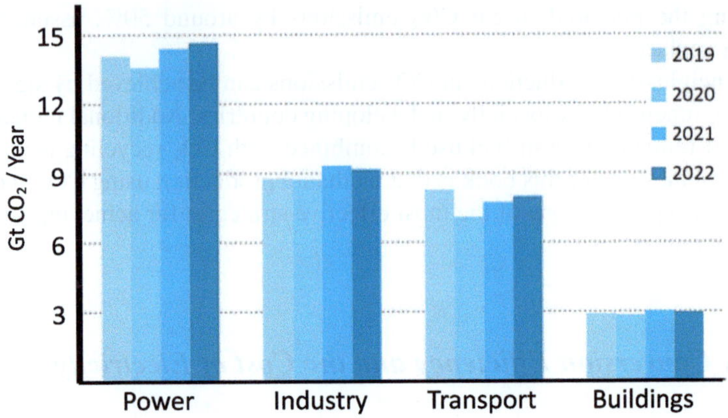

Fig. 2.3 Global CO_2 emissions by sector, 2019–2022 [5]. Data source; IEA report 2023, Reprinted from https://www.iea.org/reports/CO 2 -emissions-in-2022 under the Creative Commons Attribution 4.0 International License

accelerate this transition effectively. To address this, a detailed and quantitative discussion is needed. This chapter highlights two major challenges in the adoption of renewable energy: (1) balancing power supply and demand in the grid and (2) replacing energy provided by liquid fossil fuels such as gasoline, diesel, and heavy oil. As shown in Fig. 2.3, global CO_2 emissions are broadly categorized into the power, industry, transport, and buildings sectors. The power sector accounts for approximately 15 gigatons per year, while the combined emissions from industry, transport, and buildings reach about 20 gigatons per year. Since CO_2 emissions are directly proportional to the use of primary energy, reducing emissions requires sector-specific strategies.

Balancing Power Supply and Demand in the Grid

Solar power can provide substantial electricity during the day, but generation drops to zero after sunset. This variability poses a serious challenge, particularly in meeting demand for heating and cooling in extreme weather. Solar energy output cannot be stored directly and thus requires supplementary systems to balance supply and demand. The grid compensates for variability by using controllable sources such as thermal, hydroelectric, and pumped storage power. This concept is referred to as the simultaneity of electricity. Sources like nuclear and coal-fired power are less flexible and operate best at constant output, while gas turbines offer rapid adjustability. Fuels like LP gas are easier to control than kerosene or coal due to ignition and combustion characteristics. The so-called Duck Curve (Fig. 2.4) illustrates this problem, showing net electricity demand (excluding solar generation) throughout the day. During peak solar generation, net demand dips, but rises sharply after

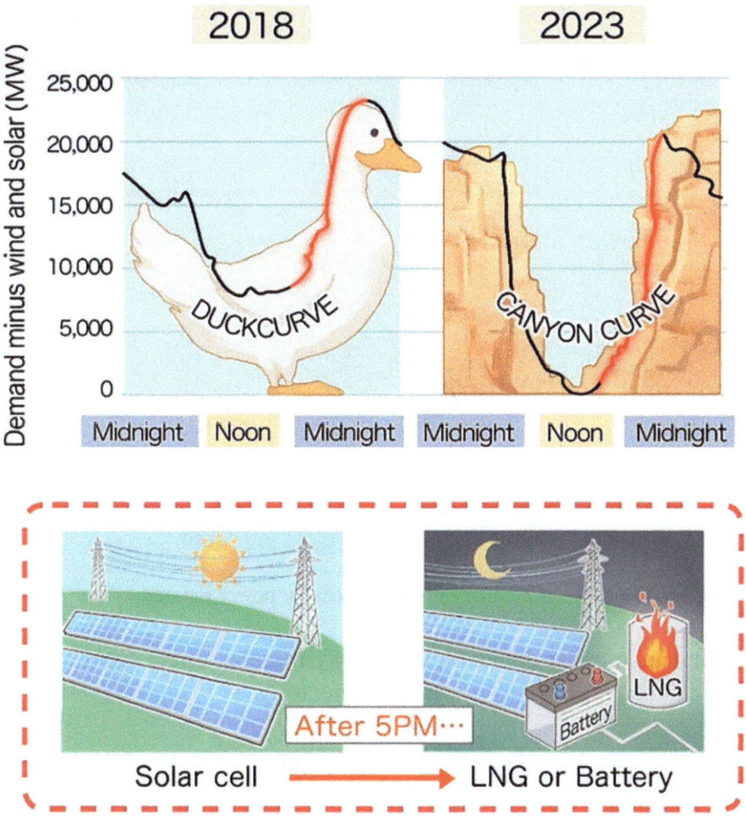

Fig. 2.4 Power demand curve excluding solar power. The situation in May 2018 (Duck curve) and April 2023 (Canyon curve)

sunset when residential and commercial usage increases. This steep evening ramp-up requires power plants to respond rapidly. Reducing power output before noon is also difficult, as gas and oil plants cannot operate below minimum capacity. In regions with high solar penetration, such as California and Hawaii, excess midday solar generation has led to over-generation issues. In some cases, utilities are forced to curtail solar input, which wastes potential energy and reduces profitability for PV producers. Without solutions to this mismatch, solar adoption may stagnate. Battery storage systems such as those in Fig. 2.5 are one proposed solution. However, large-scale deployment is limited by cost, capacity, and material constraints. For example, a large-scale plant with 444 batteries planned in Melbourne is costing US$1 billion to store 16GWh of electricity [6]. Scaling such systems is difficult due to low cost-reduction potential.

In contrast, chemical energy carriers such as hydrogen, ammonia, methanol, and methane offer advantages. They are easier to store and transport and can be synthesized using renewable electricity. These carriers can be used in thermal power or

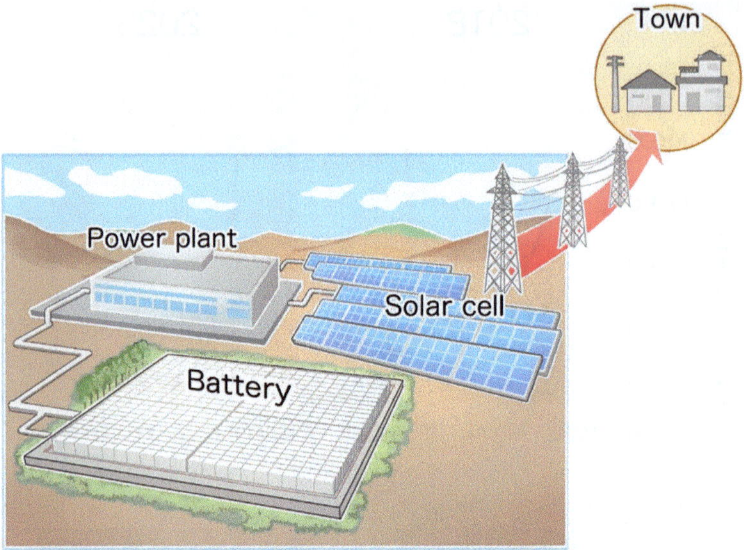

Fig. 2.5 Image of batteries as auxiliary power sources for power stations

fuel cells and are more scalable. Methanol, in particular, offers significant potential. Hydrogen can be produced from water via electrolysis:

$$H_2O \rightarrow H_2 + \frac{1}{2}O_2 \tag{2.2}$$

Then CO_2 is combined with hydrogen to produce methanol:

$$CO_2 + 3H_2 \rightarrow CH_3OH + H_2O \tag{2.3}$$

Methanol is then used as a fuel, releasing energy and emitting CO_2:

$$CH_3OH + \frac{3}{2}O_2 \rightarrow CO_2 + 2H_2O \tag{2.4}$$

This creates a closed-loop system that enables CO_2 recycling and long-term storage of renewable energy. Methanol is a liquid at ambient conditions and compatible with existing infrastructure.

Replacing Liquid Fuels in Industry, Transport, and Buildings

The second challenge is replacing fossil-based liquid fuels in the transport, industry, and buildings sectors, which together emit 20 billion tons of CO_2 annually, or 57% of total emissions. In industry, coal is used in steel and cement production, and gas or oil in chemical and textile sectors. Transport emissions come mostly from cars using gasoline or diesel, with additional emissions from ships (heavy oil) and aircraft (jet fuel). Buildings rely on gas, kerosene, and coal for heating and cooking. While electric vehicles (EVs) are expanding in the transport sector, replacing internal combustion engines with battery-powered systems presents resource challenges. Manufacturing lithium-ion batteries requires significant amounts of lithium, cobalt, and nickel. Battery production results in CO_2 emissions ranging from approximately 100–200 kg per kWh, depending on the specific energy mix, manufacturing processes, and supply chain conditions. Using a typical value of 150 kg/kWh, replacing 1.5 billion vehicles with 60 kWh batteries would result in approximately 1.35 billion tons of CO_2 emissions. In addition, EV batteries are heavy—averaging 300–400 kg compared to 100–300 kg for gasoline systems—reducing vehicle efficiency. Recycling challenges and high infrastructure costs further hinder scalability. Figure 2.6 shows the economies of scale in cost. When using energy carriers such as hydrogen or methanol, there are benefits to scaling up. Conversely, for batteries, the number of batteries needs to increase to scale up, and there are no significant economies of scale. Meanwhile, more importantly, the depletion of key materials like Li and Co presents a long-term sustainability issue. Thus, batteries alone cannot decarbonize other sectors such as iron and steel, cement, chemicals, and buildings. The only scalable alternative is chemical energy carriers derived from renewable sources. This is a key message of this book: the future of the energy transition lies in selecting and scaling the right energy carriers.

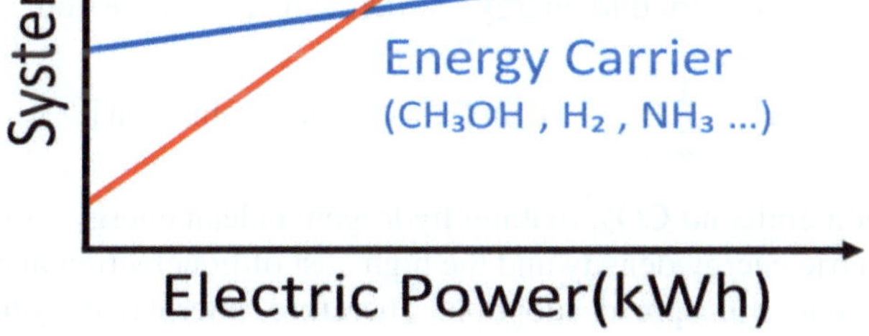

Fig. 2.6 General relationship between system cost of renewable energy conversion and electric power

2.3 Challenges and Limitations of Energy Carriers

As discussed in the previous section, energy carriers are essential for building a renewable energy-based society. There are two main methods for storing, transporting, and supplying energy derived from renewables:

i. Electricity, which is transmitted through power grids or stored in batteries.
ii. Chemical energy carriers, which are substances that can replace fossil fuels.

While electricity is a convenient and versatile form of energy, it is inherently affected by natural fluctuations. Solar power can only be generated during daylight hours, and wind power is intermittent. Storing electricity in batteries is possible, but high costs and resource constraints make large-scale global deployment infeasible. This issue will be further examined later. In contrast, storing renewable energy in the form of chemical substances—similar to fossil fuels such as natural gas, oil, and coal—offers a more scalable and versatile alternative. These chemical energy carriers must be capable of being traded internationally to meaningfully mitigate global warming. They serve multiple functions: storing and transporting energy, stabilizing renewable energy supply, and enabling decarbonization in sectors that are difficult to electrify, such as heavy industry, long-distance transport, and residential heating.

Overview of Energy Carrier Candidates

Hydrogen can be produced via water electrolysis using electricity from renewable sources:

$$H_2O \rightarrow H_2 + \frac{1}{2}O_2 \,(\text{Electrolysis}) \tag{2.5}$$

It can then be reconverted to energy via fuel cells or combustion:

$$H_2 + \frac{1}{2}O_2 \rightarrow H_2O\,(\text{Fuel cell or combustion}) \tag{2.6}$$

This reaction emits no CO_2, making hydrogen a clean energy carrier. However, its low volumetric energy density and the high cost of liquefaction and storage pose serious barriers to widespread adoption. Therefore, alternative hydrogen carriers that store H_2 in chemical form are being actively explored.

Ammonia (NH_3)

Ammonia is synthesized from hydrogen and nitrogen via the Haber-Bosch process:

$$N_2 + 3H_2 \rightarrow 2NH_3 \text{ (Fe – based catalyst)} \quad (2.7)$$

It can be decomposed to regenerate hydrogen:

$$2NH_3 \rightarrow N_2 + 3H_2 \text{ (Ru – or Ni – based catalyst)} \quad (2.8)$$

Ammonia is easily liquefied and transported and is already synthesized on an industrial scale—about 200 million tons annually (IFA, 2022). Nitrogen is obtained from air separation.

Methylcyclohexane (MCH)

MCH is produced by hydrogenating toluene:

$$C_7H_8 + 3H_2 \rightarrow C_7H_{14} \text{ (Pd – based catalyst)} \quad (2.9)$$

It can be dehydrogenated back to toluene to release hydrogen:

$$C_7H_{14} \rightarrow C_7H_8 + 3H_2 \text{ (Pt – or Pd – based catalyst)} \quad (2.10)$$

Toluene is derived from petroleum refining and coke oven processes.

Methane (CH_4)

Methane, the main component of natural gas, can be synthesized from CO_2 and H_2:

$$CO_2 + 4H_2 \rightarrow CH_4 + 2H_2O \text{ (Ni – based catalyst)} \quad (2.11)$$

It can also be reformed back into hydrogen:

$$CH_4 + 2H_2O \rightarrow CO_2 + 4H_2 \text{ (Ni – based catalyst)} \quad (2.12)$$

Methanation from CO_2 typically takes place in two steps:

$$CO_2 + H_2 \rightarrow CO + H_2O \text{ (Reverse water – gas shift)} \quad (2.13)$$

$$CO + 3H_2 \rightarrow CH_4 + H_2O \quad (2.14)$$

This process is compatible with existing LNG infrastructure, making methane a practical hydrogen carrier.

Methanol (CH_3OH)

Methanol can be synthesized from CO_2 and H_2 using Cu/ZnO catalysts:

$$CO_2 + 3H_2 \rightarrow CH_3OH + H_2O \tag{2.15}$$

Steam reforming of methanol produces hydrogen:

$$CH_3OH + H_2O \rightarrow CO_2 + 3H_2 \ (\text{Methanol steam reforming}) \tag{2.16}$$

Methanol is also produced industrially from CO and H_2. Operating conditions are typically 30–50 atm and 250 °C for synthesis and 200–300 °C for reforming.

Formic Acid (HCOOH)

Formic can be synthesized electrochemically from CO_2 and H_2O using carbon-based catalysts:

$$CO_2 + 2H^+ + 2e^- \rightarrow HCOOH \tag{2.17}$$

$$H_2O \rightarrow 2H^+ + \frac{1}{2}O_2 + 2e^- \tag{2.18}$$

Formic acid decomposes at 200–300 °C to release hydrogen:

$$HCOOH \rightarrow CO_2 + H_2 \tag{2.19}$$

This pathway avoids the need to electrolyze water directly for hydrogen, offering an alternative hydrogen storage route.

In summary, while hydrogen is a clean and flexible energy carrier, the use of chemical hydrogen carriers such as ammonia, MCH, methane, methanol, and formic acid may provide more practical solutions for global-scale energy storage and transport. These carriers can help address the intermittency of renewables and facilitate deep decarbonization across various sectors.

Evaluating Energy Carriers: A Quantitative Perspective

To assess the practicality of various energy carriers, one of the most important indicators is volumetric energy density—the amount of energy per unit volume. This is critical when considering how much energy a tanker or fuel tank can transport. A carrier with low volumetric energy density requires larger containers, reducing transport efficiency and increasing land use, construction, and maintenance costs.

As shown in Table 2.1, the volumetric energy densities of various carriers are as follows:

- Liquid hydrogen: 10.04 GJ/m^3
- Methylcyclohexane (MCH): 6.72 GJ/m^3
- Methanol: 17.93 GJ/m^3
- Ammonia: 14.35 GJ/m^3
- Crude oil: 32.28 GJ/m^3
- LNG (Liquefied Natural Gas): 25.12 GJ/m^3

Clearly, crude oil and LNG outperform alternative carriers in energy density, which is one reason they currently dominate the global energy economy. In contrast, methanol offers a practical balance: its volumetric energy density is higher than that of hydrogen, ammonia, and MCH, and it can be stored and transported in conventional containers such as glass bottles and polyethylene tanks. Although pure methanol is corrosive to iron, appropriate materials can be used to mitigate this issue. When diluted in water, methanol's energy content decreases, but safety increases. Hydrogen, while lightweight (70.8 kg/m^3), has extremely low volumetric energy density, which presents a major challenge for storage and transport. Ultimately, in transport scenarios, volume—not mass—is the limiting factor, and large volumes also increase container weight and cost. If we compare energy carriers solely by volumetric energy density, ammonia, methane, and methanol emerge as viable candidates. Liquid hydrogen and MCH are significantly less favorable in this respect. Of course, selecting an optimal energy carrier requires a broader evaluation, including:

- Cost
- Energy conversion efficiency

Table 2.1 Characteristics of energy carrier

Energy carrier	Density (kg/m^3)	Combustion heat (GJ/m^3)	Ideal conversion power (GJ/m^3)	Remarks
Crude oil $(CH_2)_n$	850.0	32.28	–	Handled at 1 atm and RT
LNG	460.0	25.12	1.52	Boiling point: −161.5 °C
Liquid H_2	70.8	10.04	1.02	−253 °C
NH_3	639.0	14.35 16.0 ($3H_2/2$)	30 MPa:1.75 20 MPa:1.51 10 MPa:1.15	P(vapor) of 1.01 MPa at 25.7 °C −33.4 °C(1 atm) Inhalation lethal dose:1500 ppm
Methyl Cyclohexane (MCH)	770.0	6.72 ($3H_2$)	1.61 (350 °C)	Handled at 1 atm and RT
CH_3OH	792.0	17.93 14.65 (with reforming)	5 MPa:1.07 1 MPa:0.68	Handled at 1 atm and RT Oral lethal dose:0.3–1 g/kg

- Toxicity and safety
- Compatibility with existing infrastructure
- Feasibility of global-scale distribution

These are the topics that will be quantitatively examined in the following chapters.

Estimating Global Energy Carrier Demand

To illustrate the magnitude of the challenge, consider that 37 billion tons of CO_2 are currently emitted annually. Replacing the associated fossil energy with synthetic carriers would require handling materials on a comparable mass scale. Depending on the carrier, this could mean producing and transporting tens of billions of tons of ammonia, methanol, or MCH. Even hydrogen, despite its low mass, would need to be produced at billions of tons per year. By comparison, current global ammonia production is about 200 million tons/year (2024). To rely on ammonia as a major energy carrier, we would need to scale this up by a factor of 100. Although the United States and European Union have pledged to produce 10–20 million tons of hydrogen annually, these targets fall far short of what is needed for carbon neutrality. This kind of quantitative reasoning is often missing in current public discourse on the energy transition. It is essential that we incorporate realistic, data-driven assessments if we are to develop viable decarbonization strategies.

Addressing Toxicity and Safety

If energy carriers are to be produced and distributed globally at massive scales, toxicity and safety become critical issues. Ammonia is a toxic and corrosive gas with a strong, irritating odor. Acute exposure to concentrations above ~1500 ppm can cause severe respiratory distress, including lung and airway damage. Lethal effects in animal studies (LC_{50} for rats, 1 h exposure) have been reported at concentrations around 2000–5000 ppm, depending on species and exposure conditions. The Immediately Dangerous to Life or Health (IDLH) concentration defined by NIOSH is 300 ppm. Ammonia is also highly toxic to aquatic life and can disrupt ecosystems by alkalinizing soil and contaminating water bodies. Large-scale ammonia spills or leaks could cause significant environmental harm. Therefore, while ammonia is an effective hydrogen carrier, its production, storage, and transport at global scale require strict safety protocols and robust risk management. While it is less acutely irritating than ammonia, methanol is highly toxic if ingested, inhaled, or absorbed through the skin as described in Sect. 5.2.

In conclusion, when evaluating energy carriers, we must weigh their technical performance against environmental and social risks. The choice of a central energy carrier for a carbon-neutral future will depend on achieving an optimal balance among energy density, safety, cost, and scalability.

2.4 Problems of Liquid Hydrogen as an Energy Carrier

One indicator for evaluating various energy carriers is the relative value of the energy required to produce the energy carrier compared to the energy inherent in the energy carrier. The energy inherent in energy carriers can be compared using heat of combustion. Meanwhile, the energy required for production includes various factors, such as compression, liquefaction, chemical reactions, and energy conversion processes. Generally, as the number of steps in a production process increases, energy losses increase. The selection of energy carriers will become increasingly important in the future, requiring detailed consideration.

Energy Efficiency and Precious Metal Catalysts in Electrolysis

Before examining the challenges specific to liquid hydrogen, it is important to address a fundamental issue with hydrogen itself: the high cost of production. When hydrogen is produced using renewable electricity, the primary method is water electrolysis:

$$H_2O \rightarrow H_2 + \frac{1}{2}O_2 \quad (\text{Electrolysis}) \tag{2.20}$$

When using hydrogen as an energy carrier, a key issue is how efficiently renewable electricity can be converted into a chemical form. The energy efficiency of electrolysis—measured by comparing the combustion enthalpy of hydrogen to the input electricity—is approximately 70%. The remaining 30% of energy loss stems from overpotentials required to overcome activation energy and mass transfer resistance. While 70% energy efficiency is not low, energy loss must be minimized as much as possible. This energy loss is reflected in the price of hydrogen. Various electrolysis methods exist—alkaline, proton exchange membrane (PEM), and high-temperature solid oxide—but all require expensive catalysts such as Pt, Ir, or Ru, especially at the oxygen-evolution electrode. High-temperature electrolysis offers slightly better efficiency (up to ~80%) but increases system costs due to the expensive membranes. The widespread adoption of electrolysis equipment requires the development of inexpensive electrocatalysts and membranes, but their commercialization will take time.

The Low Volumetric Energy Density

Hydrogen is an exceptionally clean energy carrier, but its low volumetric energy density as a gas poses a serious challenge. To address this, hydrogen is often liquefied to improve transport and storage. Let's examine the energy cost of this process. At 25 °C and 1 atm, hydrogen has a gas density of 0.08376 kg/m^3. To become a liquid, it must be cooled to −252.87 °C (20.28 K), reaching a liquid density of 70.85 kg/m^3—a volume reduction of 1/846. The energy efficiency of liquefaction is evaluated using the coefficient of performance (COP) of the cooling cycle, which for an ideal refrigerator is given by:

$$\mathrm{COP} = \frac{T_\mathrm{L}}{T_\mathrm{H} - T_\mathrm{L}} \tag{2.21}$$

where

- T_L is the absolute temperature of the low-temperature reservoir (K)
- T_H is the absolute temperature of the high-temperature source (K)

Using $T_\mathrm{H} = 298.15\,\mathrm{K}(-25°\mathrm{C})$ and $T_\mathrm{L} = 20.28\,\mathrm{K}(-252.87°\mathrm{C})$:

$$\mathrm{COP} = \frac{20.28}{298.15 - 20.28} = 0.073 \tag{2.22}$$

Intuitive Explanation: COP measures how much heat can be removed from the cold side per unit of work input. Unlike a typical "efficiency," which cannot exceed 100%, a refrigerator's COP can be greater than 1 because it moves existing heat rather than generating it. For example, a household refrigerator might remove 300 W of heat using only 100 W of electricity, giving a COP of 3. However, COP drops sharply as the temperature difference between hot and cold sides increases. For hydrogen liquefaction, the cold side is only 20 K above absolute zero, resulting in an ideal COP of just 0.073—about 7% of a domestic refrigerator's performance. In simple terms, removing the last traces of heat to reach cryogenic temperatures is vastly harder than cooling your kitchen fridge by a few degrees.

The latent heat of hydrogen vaporization is 445.4 kJ/kg, so the energy input per kg of liquefied H_2 is:

$$\text{Power input} = \frac{\text{Removed heat}}{\mathrm{COP}} = \frac{445.4}{0.073} \approx 6101\,\mathrm{kJ/kg} \tag{2.23}$$

This COP value is very low compared to conventional systems:

- Air conditioners: COP = 2.5–4
- Industrial refrigerators: COP = 1.5–5

Adding the cooling from room temperature to liquefaction temperature, the average COP is about 0.4. Given the specific heat of hydrogen gas (14.3 kJ/kg·K), cooling from 298 K to 20.28 K requires approximately 3977 kJ/kg. At COP = 0.4, this adds 9943 kJ/kg of input energy.

Thus, the total energy input required for liquefaction is:

- Cooling: 9943 kJ/kg
- Liquefaction: 6101 kJ/kg
- Total: ~16,044 kJ/kg (4.46 kWh/kg)

The minimum theoretical energy for liquefaction is estimated to be 3.94 kWh/kg (from 300 K, 1 atm) or 2.86 kWh/kg (from 300 K, 21 atm), depending on the initial pressure [7]. These differences are due to the nonlinear behavior of hydrogen's specific heat at low temperatures. Considering that the combustion enthalpy of hydrogen is 142 MJ/kg (39.3 kWh/kg), even under ideal conditions, 10% of its energy is lost during liquefaction. In practice, liquefaction consumes 2–3 times more energy than the theoretical minimum [8]. Current estimates are:

- Present: 10 kWh/kg
- Future target: 7 kWh/kg

This yields a liquefaction efficiency of 75% (current) or 82% (future).

Accounting for Primary Energy

All these calculations are based on electricity input (secondary energy). However, when considering primary energy, the situation worsens. If electricity is generated from fossil fuels with an efficiency of 50%, then only 49% of hydrogen's energy is available after liquefaction. Further losses occur during conversion back to electricity via fuel cells, often reducing the net efficiency to 30–35%. In short, liquefied hydrogen is highly inefficient from a full energy chain perspective and thus incompatible with the goals of an SDG-aligned energy system.

Technological Limitations of Hydrogen Liquefaction

Hydrogen liquefaction typically uses Claude or Linde cycles:

- Claude cycle: Uses external refrigerants.
- Linde cycle: Uses the compressed gas itself.

These technologies are mature, and significant efficiency improvements are unlikely. Some researchers propose alternative methods like magnetic refrigeration, which do not rely on the Joule-Thomson effect, but such technologies are still in early stages. A critical consideration is that most efficiency assessments are based on input electricity (secondary energy). When evaluating hydrogen in terms of resource use (primary energy), the losses double if electricity is generated at 50%

efficiency. While some argue that using renewable or nuclear energy eliminates this issue, this view ignores the true resource cost and opportunity cost of using such energy inefficiently. In conclusion, despite its clean-burning nature, liquid hydrogen has severe limitations in terms of energy efficiency, system cost, and infrastructure demand, making it a problematic choice for large-scale energy storage and transport in a decarbonized economy.

2.5 Problems with Ammonia Carriers

To reiterate, if the energy required to produce an energy carrier exceeds the energy required for the energy carrier, expressed as heat of combustion, it no longer qualifies as an energy carrier in a renewable-energy-based circular society. This section discusses the extremely high energy required to produce ammonia.

Ammonia plays an essential role as a raw material for fertilizers and thereby supports global food production. It is also used in the manufacture of explosives. As a result, industrial-scale systems for ammonia production, storage, and distribution are already well established and widely accepted in society. Consequently, ammonia has recently gained attention as a potential hydrogen carrier, synthesized via the reaction:

$$N_2 + 3H_2 \rightarrow 2NH_3 \tag{2.24}$$

using hydrogen produced by water electrolysis and nitrogen separated from the air. It can, in principle, be converted back to hydrogen at the point of use via the reverse reaction. Because ammonia is a liquid under relatively mild conditions and its handling technologies (e.g., tanks and pipelines) are mature, it appears to be a practical hydrogen carrier.

However, at present, ammonia cannot be evaluated favorably as a hydrogen carrier due to its high energy requirements during synthesis. This issue is well recognized [9]. Global ammonia production in 2020 was approximately 185 million tons, but it accounted for around 2% (8.6 EJ) of global final energy consumption and about 1.3% (450 million tons) of energy-related CO_2 emissions [10]. A significant portion of ammonia's energy content is effectively lost just in compressing N_2 and H_2 to several hundred atmospheres, as well as in separating N_2 from the air—both of which are energy-intensive steps. This presents a fundamental drawback. The additional energy input required to produce hydrogen raises doubts about its viability as an energy carrier.

The most widely adopted method for ammonia production from N_2 and H_2 is the Haber-Bosch process. In the Haber-Bosch process, nitrogen and hydrogen react in the presence of a K/Fe/Al_2O_3 catalyst under high-temperature (400–500 °C) and high-pressure (100–300 atm) conditions to produce ammonia. The catalyst serves to increase the reaction rate and lower the apparent activation energy. However, generating and maintaining these extreme conditions requires substantial energy input, typically in the form of fossil fuels and electricity. Here, we first assume that H_2 and N_2 are already obtained and estimate the energy required for ammonia synthesis alone.

A thermodynamic calculation based on ideal adiabatic compression shows that compressing 1 ton of an N_2/H_2 mixture (1:3 molar ratio) from 25 °C at 1 atm to 300 atm consumes approximately 3.84 GJ. The resulting gas temperature rises to 1914 K under ideal adiabatic compression without intercooling, so cooling is required to maintain the reaction temperature at 400–500 °C.

This value aligns with current literature. Given a typical ammonia yield of 80% and an energy efficiency of 50%, about 10 GJ of energy is required to produce 1 ton of ammonia. This figure should be regarded as an upper-bound estimate, consistent with values reported in the literature:

$$3.84\text{GJ} \times \frac{1}{0.8} \times \frac{1}{0.5} = 9.6\text{GJ}$$

Here, the lower heating value (LHV) represents the usable energy content of the energy carrier, expressed as the heat released during complete combustion without recovering the latent heat of vaporization of water produced. For ammonia, the LHV is typically expressed at 15 °C and includes the latent heat associated with phase change under these conditions. The LHV of 1 ton of ammonia is approximately 22.5 GJ, so 44.4% of the LHV is lost in the synthesis of ammonia alone. We will show energy estimations for hydrogen production and nitrogen separation later, but since the total energy required to produce ammonia exceeds the LHV of ammonia, ammonia cannot be used as an energy carrier in a renewable-energy-based circular society.

The basic equations for adiabatic compression are shown below, and readers are encouraged to use them for their own calculations:

Temperature Change Under Adiabatic Expansion/Compression

$$\Delta T = \frac{\Delta U}{nC_V} = T_i\left(\alpha^{\gamma-1} - 1\right) = T_i\left(\beta^{\frac{\gamma-1}{\gamma}} - 1\right) \tag{2.25}$$

T_i: Initial temperature (K), $\alpha\left(= \frac{V_i}{V_f}\right)$: Compression ratio, $\beta\left(= \frac{p_f}{p_i}\right)$: Pressure ratio,

γ: Specific heat ratio, $\Delta T = T_f - T_i$, $\alpha = \beta^{\frac{1}{\gamma}}$, T_f: Final temperature (K).

Work Done Due to Thermal Changes

$$W = nC_v\left(T_i - T_f\right) = \frac{nRT_i}{\gamma - 1}\left\{1 - \left(\frac{V_i}{V_f}\right)^{\gamma-1}\right\} \tag{2.26}$$

Unfortunately, additional energy losses arise when using ammonia as a hydrogen carrier due to the following factors:

- Electrical energy required to separate nitrogen from air.
- High thermal energy needed for ammonia cracking ($2NH_3 \rightarrow N_2 + 3H_2$), which cannot be supplied by heat pumps.
- Low overall efficiency due to compounding losses across production, conversion, and transport steps.

As a result, the overall energy efficiency becomes critically low, making ammonia a less viable choice as an energy carrier under current technological and economic constraints.

Current industrial ammonia synthesis processes are primarily designed to produce ammonia as a high-value chemical product (e.g., fertilizer), not as a fuel. Historical improvements have led to the development of more energy-efficient production methods. One such method uses methane (CH_4), the main component of natural gas, and water (H_2O) as the hydrogen source as shown in Fig. 2.7. Hydrogen is produced by steam reforming, as represented by the following reaction:

$$CH_4 + H_2O \rightarrow CO + 3H_2 \tag{2.27}$$

The carbon monoxide (CO) produced is subsequently converted into carbon dioxide (CO_2) via oxidation using either oxygen from air or water-gas shift reactions:

$$CO + \frac{1}{2}O_2 \rightarrow CO_2 \tag{2.28}$$

$$CO + H_2O \rightarrow CO_2 + H_2 \tag{2.29}$$

The resulting gas mixture primarily consists of nitrogen (N_2), hydrogen (H_2), carbon dioxide (CO_2), and residual water vapor. Ammonia (NH_3) is synthesized from the N_2 and H_2 components via the Haber-Bosch reaction:

$$N_2 + 3H_2 \rightarrow 2NH_3 \tag{2.30}$$

The unreacted CO_2 in the gas stream is typically captured and sequestered using carbon capture and storage (CCS) technologies. In cases where methanol synthesis facilities are co-located, part of the CO can be diverted for methanol production:

$$CO + 2H_2 \rightarrow CH_3OH \tag{2.31}$$

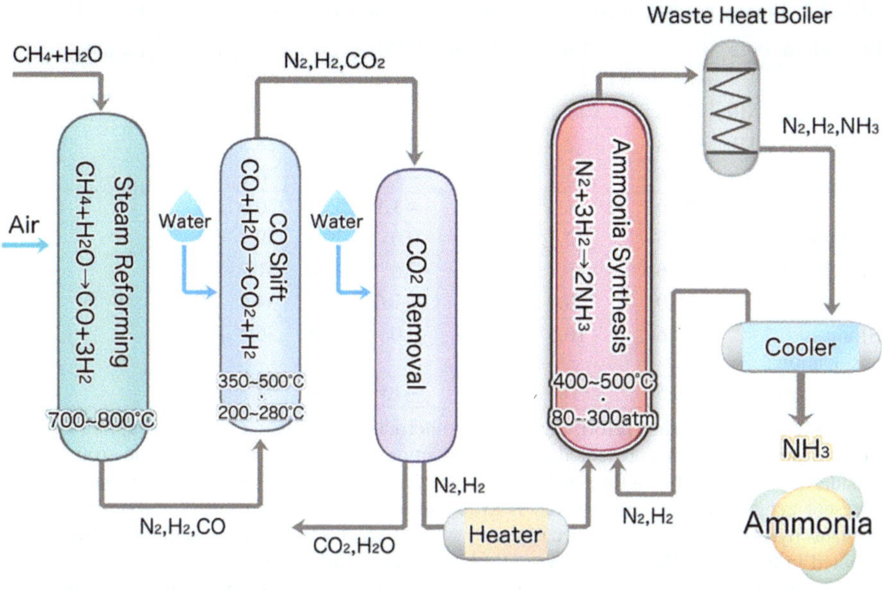

Fig. 2.7 Flow diagram of the current industrial ammonia synthesis process via the Haber-Bosch method using CH_4, H_2O, and air as feedstocks

In this process, the oxygen from ambient air is utilized directly for CO oxidation, eliminating the need for oxygen separation. The hydrogen sources—natural gas and water—are relatively inexpensive, and even though the overall process requires substantial energy input, this energy cost is generally absorbed into the market price of ammonia, which remains commercially viable due to steady demand in agriculture. The existing infrastructure for ammonia production and distribution—particularly in the agricultural sector—contributes to its high degree of social acceptance. Within the food supply chain, relatively high production costs are permissible due to the limited distribution volume. However, when considering ammonia as a large-scale energy carrier, the context changes significantly. To replace fossil fuels, the distribution volume would need to be scaled up by approximately two orders of magnitude (i.e., exceeding 20 billion tons annually). In such a scenario, energy efficiency becomes a critical requirement. If the energy efficiency of ammonia synthesis and utilization is not sufficiently high, ammonia cannot be regarded as a viable energy carrier for large-scale applications.

In a renewable-energy-based circular society, the energy required to produce an energy carrier must not exceed its lower heating value (LHV); otherwise, the carrier cannot regenerate itself without external energy input. When production energy exceeds the LHV, the cycle becomes unsustainable, as each use yields less energy than is needed for the next production cycle. As mentioned above, assuming that both nitrogen and hydrogen are readily available, the theoretical minimum energy required for ammonia synthesis is approximately 10 GJ per ton of ammonia.

Table 2.2 Energy needs to produce one tonne of ammonia for each route using BAT [11]

Production route	Energy intensity (GJ/t)						Direct CO_2 intensity
	Feedstock	Fuel	Electricity	Steam	Gross	Net	(t CO_2/t)
Natural gas SMR	21.0	11.1	0.3	−4.8	32.4	27.6	1.8
Natural gas ATR	25.8	2.1	1.0	0.0	28.9	28.9	1.6
Coal gasification	18.6	15.1	3.7	−1.3	37.4	36.1	3.2
SMR with CCS	21.0	11.1	1.0	−3.1	33.1	30.0	0.1
ATR with CCS	25.8	2.1	1.5	0.0	29.4	29.4	0.1
Coal with CCS	18.6	15.1	4.9	2.6	38.6	41.2	0.2
Electrolysis	0.0	0.0	36.0	−1.6	36.0	34.4	0.0
Biomass gasification	18.6	16.5	1.4	0.0	36.5	36.5	0.0
Methane pyrolysis	40.5	0.0	8.4	−1.6	48.9	47.3	0.0

Notes: *SMR* steam methane reforming, *ATR* autothermal reforming, *CCS* carbon capture and storage. All energy intensities presented for the production of 1 tonne of ammonia. The arrangements considered here include capture of both concentrated and dilute CO_2 streams. Negative values represent net steam generation, which is available for use by other process units or for export. Methane pyrolysis produces solid carbon as a by-product, the energy content of which is not reflected in the values shown. Methane pyrolysis is at an early stage of technology development and the estimates provided are for a future commercial-scale installation. CO_2 is generated from the use of bioenergy, but does not contribute toward direct CO_2 emissions from the energy system, following IEA emissions accounting conventions.
Source: Data gathered and reviewed in collaboration with the IFA and its members.

However, in actual industrial processes, the total energy consumption ranges from 30 to 50 GJ per ton, as shown in Table 2.2. The deviation from the ideal value of 10 GJ is primarily due to the energy-intensive process of hydrogen production. Whether natural gas, coal, or electricity is used as the primary energy source, the total energy input significantly exceeds ammonia's lower heating value (LHV) of 22.5 GJ per ton. In this situation, ammonia is unsuitable as an energy carrier. Consider the case of producing hydrogen using renewable electricity. According to Table 2.2, the total energy consumption for ammonia synthesis in this case is 34.4 GJ per ton. Unless this total energy consumption is reduced to ammonia's LHV of 22.5 GJ/ton or less, it cannot be used as an energy carrier. Even when inexpensive natural gas is used, the total energy input is 27.6 GJ/ton, making ammonia unusable.

References

1. Energy Institute (2024) 2044 Statistical review of World Energy. https://www.energyinst.org/statistical-review. Accessed 11 June 2025
2. IEA Report (2024) Chapter 2. In: Energy efficiency 2024. IEA, p 37
3. Our World in Data, Climate Change. Energy Institute—Statistical Review of World Energy (2024). https://ourworldindata.org/global-energy-200-years. Smil (2017)—with major processing by Our World in Data. Accessed 11 June 2025

4. IEA, IRENA, UNSD, World Bank, WHO (2024) P.74, tracking SDG 7: the energy progress report. World Bank, Washington DC. © World Bank. License: Creative, gjCommons Attribution—NonCommercial 3.0 IGO (CC BY-NC 3.0 IGO)
5. IEA report 2023: P.9, CO_2 Emissions in 2022, https://www.iea.org/reports/CO_2-emissions-in-2022. License: CC BY 4.0. Accessed 11 June 2025
6. https://reneweconomy.com.au/giant-melbourne-battery-may-triple-in-size-with-plans-for-12-hour-flow-battery-storage/. Accessed 11 June 2025
7. Berstat DO et al (2009) Comparison criteria for large-scale hydrogen liquefaction processes. Int J Hydrog Energy 34:1560–1568
8. Yamashita S, Shingo M, Unno S, Imai T, Sugawa S (2016) Development of hydrogen liquefaction and storage system. J Jpn Soc Energy Resour 37(4):1–8
9. Rouwenhorst KHR, Travis AS, Lefferts L (2022) 1921–2021: a century of renewable ammonia synthesis. Sustain Chem 3:149–171
10. IEA Reports (2021) Ammonia technology roadmap towards more sustainable nitrogen fertiliser production, License: CC BY-NC 3.0 IGO

Chapter 3
Issues in the Industrial Sector

Abstract This chapter addresses CO_2 emissions from key industrial sectors—steel, cement, and chemicals—and their substantial contribution to global warming. Collectively, these sectors account for approximately 25% of total global CO_2 emissions, making the industrial domain a critical focus for climate mitigation efforts. This chapter begins with an analysis of the steel industry, where the use of coke for iron ore reduction is a major source of emissions. It then examines the cement industry, highlighting both combustion-related emissions and those from limestone calcination. The chemical industry's dependence on fossil fuels as both energy and feedstocks is also scrutinized. Additionally, this chapter considers the growing problem of plastic waste and its environmental consequences. Each sector is evaluated in terms of emission sources, technological constraints, and the limited potential for electrification. This chapter concludes by arguing that scalable chemical solutions—such as converting captured CO_2 into methanol—are essential for meaningful decarbonization in these hard-to-abate sectors.

Keywords Industrial emissions · CO_2 reduction · Steel industry · Cement production · Chemical processes · Plastic waste · Methanol pathway

3.1 Challenges in the Steel Industry

The steel industry is a major contributor to global CO_2 emissions, accounting for approximately 3.6 gigatons annually, or about 7–10% of the global total [1]. Some estimates report a lower figure of 2.6 gigatons [2], depending on whether emissions from electric arc furnaces (EAFs) are included. Regardless of the exact number, decarbonizing the steel sector and deploying carbon capture, utilization, and storage (CCUS) technologies are essential for global climate change mitigation.

In certain regions such as Japan and the EU, efforts toward carbon neutrality in the steel industry by 2050 have been underway since around 2009. Research and development in the field have accelerated, and CCUS initiatives have commenced. As of 2024, 45 commercial facilities are operating globally, capturing a total of

J. Nakamura, M. Ishida, *Blueprint for a Methanol Society*,
https://doi.org/10.1007/978-981-95-3465-4_3

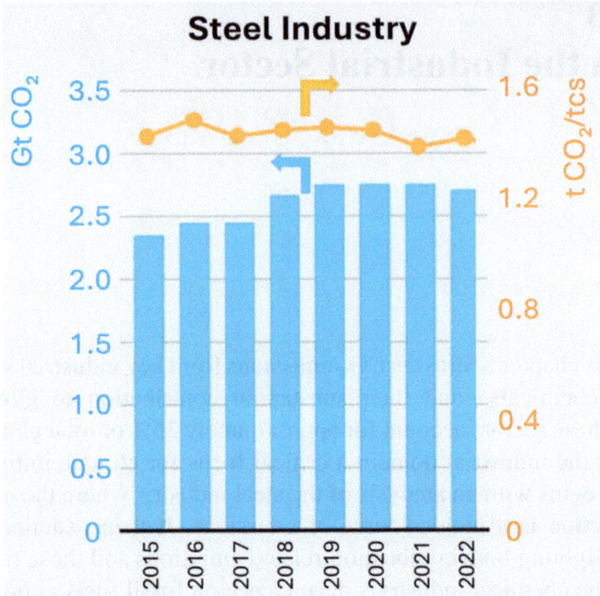

Fig. 3.1 Annual CO_2 emission from steel Industry. Adapted from [2] with permission by the International Energy Agency (IEA)

approximately 50 million tons of CO_2 per year [3]. However, this is far from sufficient to achieve carbon neutrality. As shown in Fig. 3.1, CO_2 emissions from the steel industry have not significantly declined. While emissions per ton of steel have slightly decreased since 2016, the overall reduction is inadequate. Despite optimistic projections targeting 2030, current trends indicate little progress. This highlights a critical issue: merely setting targets does not guarantee their realization.

The fundamental challenge lies in replacing fossil fuels, which have historically been low cost, with alternative energy sources. In this section, we examine the present state of CO_2 emissions in steelmaking and assess various strategies for reduction.

CO_2 Emissions in the Blast Furnace Process

Steel is primarily produced using three methods: the blast furnace (BF) process, the electric arc furnace (EAF) process, and direct reduced iron (DRI) methods. Among these, the traditional BF process accounts for the majority of global iron production. In a BF, alternating layers of iron ore (Fe_2O_3), as shown in Fig. 3.2, and coke are charged, and hot air is injected from the bottom. Pig iron is produced via the following reduction reaction:

$$Fe_2O_3 + 3CO \rightarrow 2Fe + 3CO_2 \tag{3.1}$$

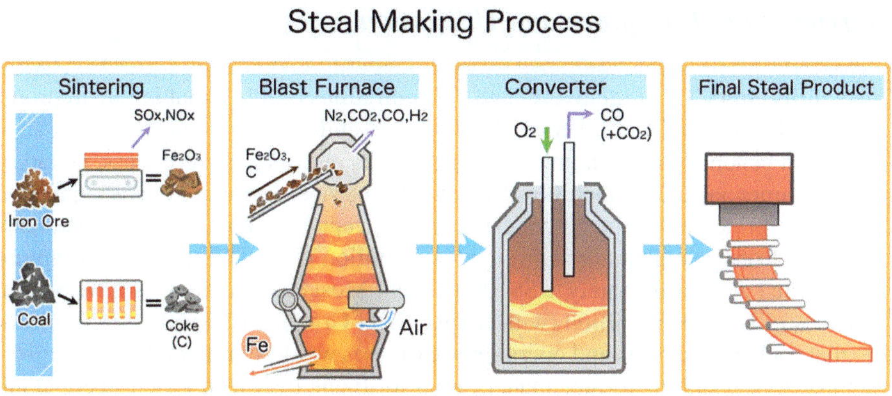

Fig. 3.2 Steal making process from iron ore and coal

The resulting pig iron is transferred to a basic oxygen furnace, where impurities such as carbon and sulfur are removed through combustion:

$$C + O_2 \rightarrow CO_2 \tag{3.2}$$

Theoretically, based on the molar masses of Fe (55.845 g/mol) and CO_2 (44.01 g/mol), approximately 1.18 tons of CO_2 are generated per ton of iron. However, in practice, the emission is around 2.0–2.2 tons of CO_2 per ton of crude steel. Global crude steel production reached 1.885 billion tons in 2023.

The International Energy Agency (IEA) projects that by 2050, 400 million tons of CO_2 could be captured annually in the steel industry. Nonetheless, this remains well below the threshold required for carbon neutrality.

Technological Pathways for Decarbonization

Efforts to reduce CO_2 emissions in the steel sector include hydrogen-based direct reduction, EAF steelmaking, and CCUS. Among these, hydrogen-based steelmaking is receiving considerable attention. In this approach, hydrogen (H_2) replaces carbon monoxide as the reducing agent, and water (H_2O) is produced instead of CO_2. Although promising, the technology is currently in the demonstration phase. Major obstacles include the high capital investment and the cost of hydrogen. In addition, hydrogen storage and transport present significant logistical challenges. The infrastructure costs make it difficult for widespread adoption outside developed nations. Moreover, the liquefaction of hydrogen reduces energy efficiency and increases costs.

Barriers to CO_2 Capture and Utilization

Steel mills emit CO_2 at relatively high concentrations (~20%), making them promising candidates for CCUS. Unlike the dispersed emissions from automobiles, steel plants can benefit from economies of scale in CO_2 recovery. Nevertheless, no large-scale commercial CCUS facilities dedicated to the steel sector are operational as of 2024. The principal barrier to widespread CO_2 capture is cost. Initial investments can range from several hundred million to one billion USD. The cost of CO_2 capture is estimated at approximately US$100 per ton. Most steel companies have annual revenues of US$10–60 billion (with Baowu Steel Group at US$160 billion), meaning that such capital costs and operational expenses would significantly impact profitability. As a result, many steelmakers are prioritizing hydrogen-based steelmaking over CO_2 capture [4]. Substantial public-sector support will be essential for CO_2 capture to advance.

Another hindrance is the limited development of CO_2 utilization. Carbon capture and storage (CCS), unlike CCUS, involves injecting captured CO_2 underground without reuse. Although this prevents atmospheric emissions, the additional storage costs and lack of economic return make it less attractive. IEA sees hydrogen-based steelmaking as a key long-term pathway, while also emphasizing the role of CCUS [4].

Despite growing interest in hydrogen, the steel sector still lacks a comprehensive energy carrier transition strategy. Even if hydrogen-based methods gain traction, their global dissemination—especially in developing countries—will be limited in the short to medium term. While advancing hydrogen-based steelmaking is not discouraged, CO_2 capture remains the most immediate and scalable mitigation option. Its neglect on the grounds of cost alone is unconvincing. Political intervention is necessary.

Toward a Methanol-Based Solution

Captured CO_2 can be converted into methanol (CH_3OH), which serves as both a fuel and a chemical feedstock. Near steel plants, methanol can be reconverted to CO_2 and H_2, enabling integration with hydrogen-based steelmaking. Alternatively, methanol can be used directly as a fuel. These approaches offer practical pathways to accelerate decarbonization in the steel industry.

CO_2 should be regarded as a valuable carbon resource. In nature, photosynthesis converts CO_2 and H_2O into carbohydrates and O_2, while respiration returns CO_2 and H_2O through the oxidation of organic compounds. This natural carbon cycle provides a model for synthetic analogs. In the methanol economy proposed in this volume, captured CO_2 is converted to methanol, used, and then recaptured to close the loop. Such a circular process aligns with natural systems and promotes sustainability.

Methanol can potentially be used directly in blast furnaces. As will be described in Part II, pilot-scale testing of methanol-based ironmaking is ongoing in Europe. These developments should be actively supported. Ultimately, it is ideal to install CO_2 capture and methanol synthesis systems on-site at blast furnaces, enabling integrated fuel recycling.

3.2 Challenges in the Cement Industry

The cement industry is among the largest industrial emitters of CO_2, typically ranked second after the steel industry in most estimates, releasing approximately 2.5 gigatons annually. Despite the urgent need for decarbonization, emission reduction in this sector has seen limited progress. As illustrated by the light blue bars in Fig. 3.3, emissions have been increasing since 2015, with only a slight decline observed in 2022. The total emissions remain around 2.5 Gt/year, and the outlook for significant reductions by 2030 appears bleak.

The International Energy Agency (IEA) published a road map for the cement sector in 2009, outlining a plan to reduce CO_2 emissions by 18% from 2007 levels by 2050 [6]. However, emissions have continued to rise, highlighting a critical shortfall in mitigation efforts. This issue must be addressed in conjunction with carbon neutrality measures across all sectors.

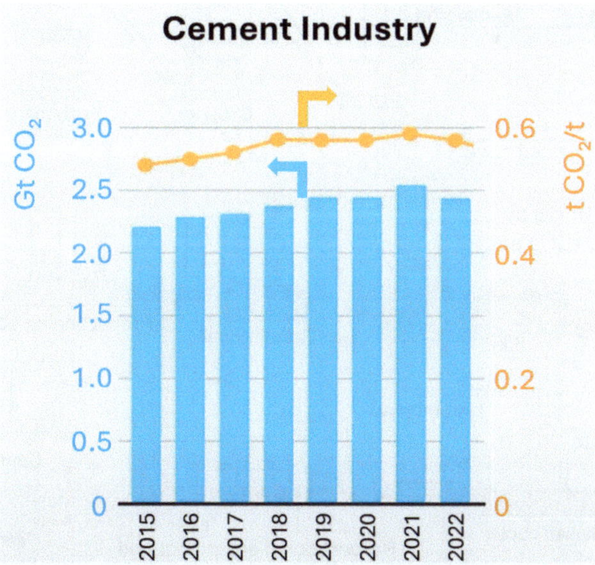

Fig. 3.3 Annual CO_2 emission from cement industry. Adapted from [5] with permission by the International Energy Agency (IEA)

Sources of CO_2 Emissions in Cement Manufacturing

Cement is manufactured primarily from limestone, as shown in Fig. 3.4, whose main component is calcium carbonate ($CaCO_3$). Other raw materials such as silicon dioxide (SiO_2), aluminum oxide (Al_2O_3), and iron oxide (Fe_2O_3) are added, and the mixture is sintered at temperatures exceeding 1450 °C using rotary kilns. Coal is the dominant fuel used to achieve these high temperatures.

During kiln operation, the raw materials react to form clinker, a complex mixture of oxides. The clinker is subsequently ground and mixed with a small amount of gypsum ($CaSO_4$) to produce cement.

A significant portion of CO_2 emissions arises from the calcination of limestone:

$$CaCO_3 \rightarrow CaO + CO_2 \tag{3.3}$$

This reaction accounts for approximately 50% of total CO_2 emissions in cement production. An additional ~40% comes from the combustion of fossil fuels, primarily coal, during clinker formation.

The flue gas from cement kilns contains a relatively high CO_2 concentration, typically in the range of 20–30%. In the preheater section, CO_2 concentrations

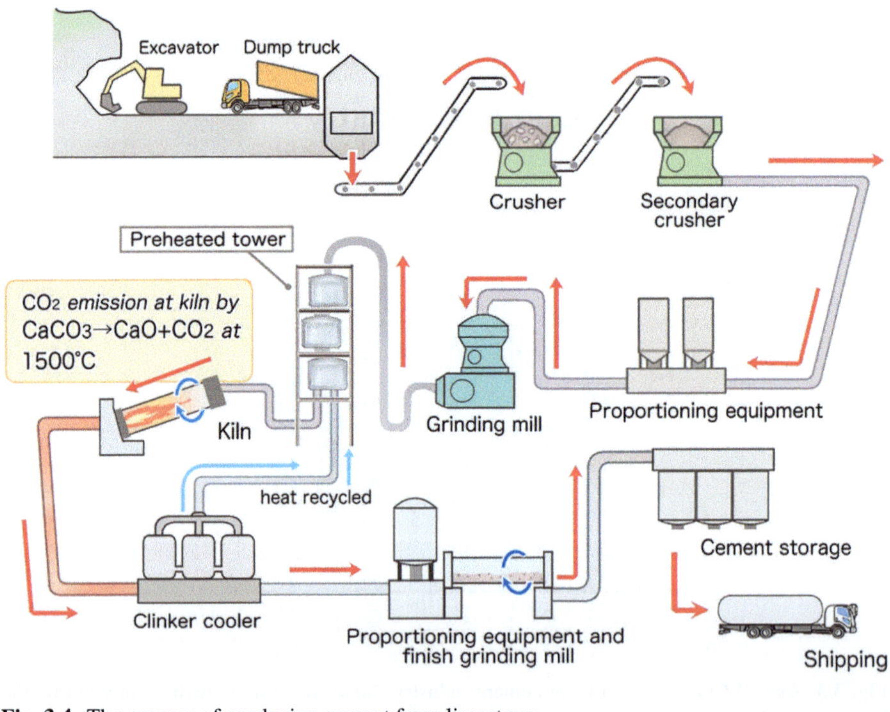

Fig. 3.4 The process of producing cement from limestone

range from 15 to 25%. These high concentrations offer favorable conditions for carbon capture with relatively high efficiency.

Current Status and Prospects for CO_2 Capture

Despite the technical feasibility, as of 2024, no cement plants have adopted CO_2 capture as a full-scale commercial process. The primary barrier remains the substantial initial investment required for carbon capture and storage (CCS) or carbon capture, utilization, and storage (CCUS) infrastructure. Several pilot and demonstration projects are underway globally.

LafargeHolcim, in collaboration with Carbon Clean Solutions, has explored pilot-scale processes to convert captured CO_2 into methanol. In Iceland, initiatives have been tested to synthesize methanol from CO_2, in some cases in cooperation with cement plants, although most projects to date use CO_2 from geothermal sources. In parallel, efforts are underway to permanently store CO_2 underground. For instance, Norcem's cement plant in Brevik, Norway, is preparing to begin CCS operations in 2024, targeting 400,000 tons of CO_2 capture per year using amine absorption technology [7]. Similarly, Lehigh Hanson's cement plant near Edmonton, Canada, aims to capture and store approximately 600,000 tons of CO_2 annually.

These projects are commendable as part of broader climate action efforts. Captured CO_2 can also serve as a feedstock for methanol synthesis. Given the large volume of CO_2 available, it is technically feasible to produce methanol at commercial scales. However, the economic challenge lies in the cost of hydrogen. In 2024, methanol prices range between 380 and 680 USD per ton. Producing one ton of methanol requires approximately 0.185 tons of green hydrogen, which currently costs around 7000 USD per ton. Thus, the hydrogen input alone costs approximately 1312 USD—more than the market price of methanol.

Despite these economic constraints, proactive steps to reduce CO_2 emissions must be taken, even at financial cost. The high capital costs of CO_2 capture systems remain the greatest barrier, and political and regulatory support will be essential to enable widespread deployment. This calls for coordinated international efforts to subsidize and accelerate decarbonization in the cement industry.

3.3 Problems in the Chemical Industry

CO_2 Emissions and the Chemical Industry

The CO_2 emissions from the global chemical industry are estimated at around 1.8 Gt per year, equivalent to about 5% of global total energy-related emissions; when including process emissions and feedstock-related CO_2, the share can reach around 10% (as of 2023) [8]. In the chemical industry, products containing carbon elements such as plastics are manufactured from fossil resources, but including the CO_2 emissions from their waste combustion, it becomes 1.8–2.0 billion tons. The global CO_2 emissions at the chemical manufacturing stage were 0.94 billion tons in 2022, as seen in Fig. 3.5 [9]. Of this, the largest is the synthesis of 0.42 billion tons of ammonia. Methanol production accounts for about 0.26 Gt of CO_2 during manufacturing, excluding downstream emissions from its use as fuel or chemical feedstock. The problem with CO_2 emissions in the chemical industry is that, in addition to the CO_2 emissions during the manufacturing process, an equivalent amount of CO_2 is emitted after the consumption of chemical products. In particular, plastics, which are mainly composed of carbon and hydrogen atoms and denoted as CHx with a molecular weight of about 14, become CO_2 with a molecular weight of 44 when they burn and combine with two oxygen atoms. In other words, plastics emit about three times their own weight in CO_2. Figure 3.5 states that the CO_2 emissions during the manufacturing process are 0.94 billion tons, but if the 0.42 billion tons of plastic produced (2020) all turns into CO_2, it becomes about 1.2 billion tons. Plastic production is expected to continue to increase in the future [10]. In the future, a CO_2 recycling society is desired, so these figures are important when planning CO_2 recovery from plastics.

Let's consider the uses of ammonia and methanol and their CO_2 emissions. Currently, the production volumes of ammonia and methanol are about 0.18 billion tons and 0.10 billion tons, respectively. Ammonia is the raw material for most

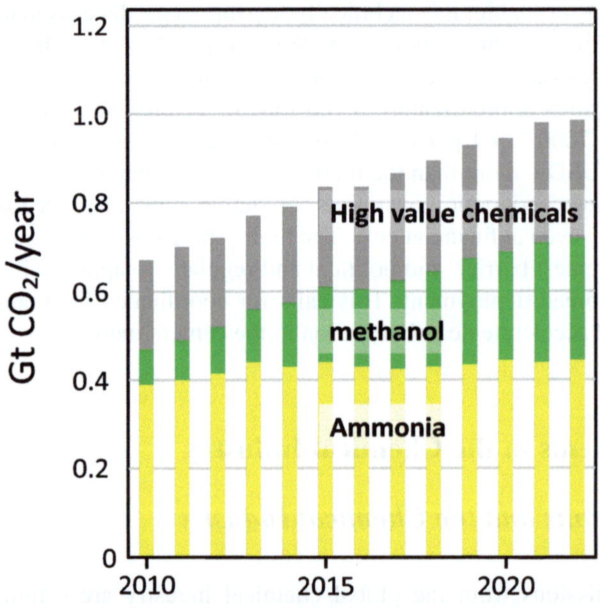

Fig. 3.5 Direct CO_2 emissions from primary chemical production, Data from [9] Data source; IEA, (Reprinted from https://www.iea.org/energy-system/industry/chemicals under the Creative Commons Attribution 4.0 International License)

nitrogen fertilizers, and half of the world's food supply depends on synthesized ammonia fertilizers. About 70% of ammonia is used for fertilizer production, and the rest is mainly used for the production of plastics, explosives, and synthetic fibers. Methanol is mainly used for the production of formaldehyde, as fuel, and as a raw material for olefin synthesis. Formaldehyde is used for the production of acetic acid, plastics, coatings, and acetic acid. In recent years, its use as fuel has been increasing. Also, ethylene, propylene, benzene, toluene, and xylene are synthesized by MTO (methanol to olefin). Plastics are manufactured from these olefins. In particular, methanol production in China has been increasing significantly. In ammonia synthesis, methane is used as a source of hydrogen, so CO_2 derived from the carbon of methane is emitted. It is said that 1.8 tons of CO_2 are emitted per ton of ammonia [10]. In the future, it is desirable to use hydrogen derived from renewable energy. Methanol is also synthesized using fossil resources such as natural gas and coal as sources of carbon and hydrogen. Since methanol contains carbon atoms, CO_2 is emitted after it is used as fuel or plastic. Ammonia and methanol are seen as promising energy carriers. There is no doubt that they can contribute to urgent carbon-neutral measures because there is accumulated manufacturing technology. However, it is necessary to consider which is more advantageous. One of the purposes of this book is to discuss that.

The Role and Challenges of the Petrochemical Industry

Let's discuss what the chemical industry is. The chemical industry generates annual revenues of US$4.7 trillion, accounting for about 4% of the world's GDP [10]. It also directly employs more than 15 million people. The core of the chemical industry is the petrochemical industry, which produces various chemical products using oil and natural gas as raw materials. Petrochemical products are essential to modern industrial society and the lives of citizens. Since oil is mainly composed of carbon and hydrogen, the main component of petrochemical products is carbon. Used petrochemical products are disposed of by waste or combustion. Petroleum- and natural gas-related chemical industry products account for approximately 50% of the chemical industry's total sales. The main petrochemical product is plastic and the raw materials are ethylene and propylene. Currently, the disposal of plastics is causing serious problems. When burned, CO_2 is released into the atmosphere. If we continue in this manner, the accumulation of CO_2 in the atmosphere will continue to increase, and waste will be discarded in the ground, transforming our planet into a garbage-filled world. This wasteful use of carbon resources is a problem. Conversely, fossil resources are important material resources for the manufacture of chemical products. For the sake of future generations, we should avoid consuming resources prematurely and use them effectively little by little. In other words, we need to seriously consider the future of the chemical industry.

Here, let's look at how and how much oil is used in the current petrochemical industry. In the second part of the Methanol Society's chemical use, we propose a society that recovers exhaust CO_2 from combustion to produce methanol and uses that methanol as a starting material to produce chemical products. To do this, it is necessary to understand the problems of the current petrochemical industry.

First, let's look at what petrochemical products are.

i. **Basic chemicals**: Hydrocarbons such as ethylene, propylene, butadiene, benzene, toluene, xylene, and aromatic compounds. These are used as raw materials for various chemical products.
ii. **Polymers**: Synthetic resins such as polyethylene, polypropylene, polyvinyl chloride (PVC), polyester, and polyurethane. These are used as raw materials for plastic products and synthetic fibers.
iii. **Fibers**: Synthetic fibers such as polyester, nylon, and acrylic. These are used in clothing and household goods.
iv. **Pesticides**: Pesticides such as pyrethroids, carbamates, and organophosphates. These are used for pest and disease control in agriculture.
v. **Pharmaceuticals**: Pharmaceutical raw materials such as aspirin and acetaminophen. These are used for medical purposes.
vi. **Synthetic Rubber**: Synthetic rubbers such as styrene-butadiene rubber (SBR), polybutadiene rubber (BR), and nitrile rubber (NBR). These are used in automobile tires and industrial products.
vii. **Synthetic Resins**: Synthetic resins such as phenolic resin, epoxy resin, and acrylic resin. These are used in adhesives, paints, and product manufacturing.
viii. **Chemical Fertilizers**: Chemical fertilizers such as nitrogen fertilizer, phosphate fertilizer, and potassium fertilizer. These are used for nutrient supplementation in crops in agriculture.
ix. **Surfactants**: Surfactants such as alkylbenzenesulfonate (ABS) and sodium lauryl sulfate. These are used in detergents and cosmetics.
x. **Additives**: Additives such as preservatives, colorants, and fragrances. These are used in food, cosmetics, pharmaceuticals, etc.

In the petrochemical industry, ethylene, propylene, benzene, xylene, etc. are produced as main products. The market size of petrochemical products is huge, reaching about 4.5 trillion dollars in 2020. The Asia-Pacific region is the largest market, and the Middle East region is also showing rapid growth. The direction of how to reduce CO_2 emissions in the petrochemical industry is not yet determined. Currently, chemical companies are at a turning point and are exploring future measures. In doing so, it is a major challenge. Through technological innovation, the petrochemical industry is aiming to improve efficiency and reduce environmental impact. By introducing new catalysts and process technologies, they are aiming to improve product quality and production efficiency. Efforts toward sustainability are progressing, such as the use of renewable energy and the introduction of biomass-derived raw materials. However, the situation is still far from carbon neutral, and there are no reports that CO_2 emissions have decreased.

The core of the petrochemical industry is plastic manufacturing, and its main raw materials are ethylene and propylene as shown in Fig. 3.6. The world production of ethylene and propylene in 2019 is about 190 million tons and about 120 million

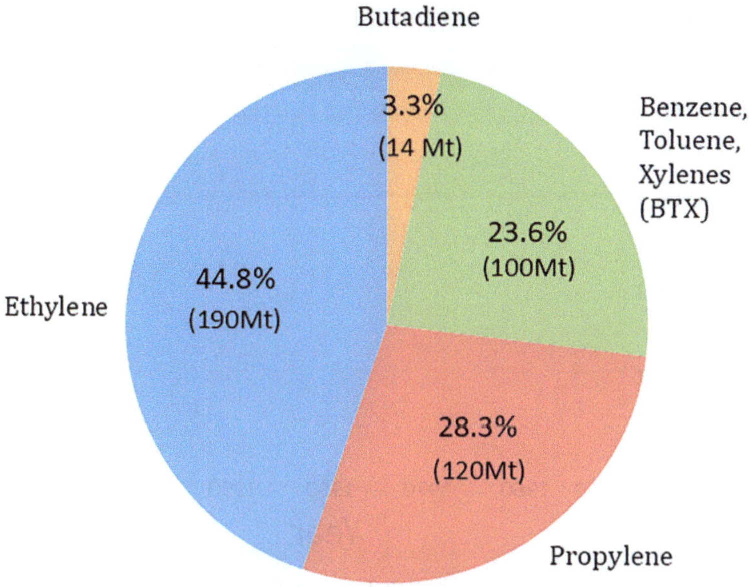

Fig. 3.6 Global production of main petrochemical products (2019, Wikipedia etc.)

tons, respectively. The annual production of plastic is over 430 million tons currently, as shown in Fig. 3.7, but the difference in weight from ethylene and propylene is because it includes other chemicals such as benzene and xylene. Benzene, xylene, and toluene are abbreviated as BTX. Polyxylene is the raw material for PET bottles. Benzene and xylene are used in polystyrene (such as Styrofoam), ABS resin, and epoxy resin. Toluene is widely used in paints, adhesives, inks, varnishes, industrial cleaning products, explosives, and dyes. Ethylene and propylene are made from naphtha. Naphtha is primarily used as a petrochemical feedstock for producing ethylene and propylene and to a lesser extent as a blending component in gasoline. Other petroleum fractions, such as kerosene, diesel, and heavy fuel oil, are derived separately in refinery processes. Naphtha is first separated by distillation. By distillation, naphtha is divided into fractions, light gasoline, kerosene, and heavy residues. The heavy residue is broken down into lighter compounds by a process called cracking. This produces important basic chemicals such as ethylene and propylene. Basic chemicals such as ethylene and propylene are converted into high molecular compounds by polymerization reactions. This produces plastics such as polyethylene and polypropylene. Various other chemicals are also made from naphtha, such as paints, synthetic rubber, pesticides, and fertilizers.

Also, about 30–40% of naphtha is used as a raw material for gasoline. Gasoline is mainly used as fuel for internal combustion engine vehicles, such as cars and motorcycles. About 20–30% of naphtha is light oil. Light oil is used as fuel for diesel engine vehicles and ships. About 10–20% of naphtha is heavy oil. Heavy oil is used as fuel for heavy oil boilers. About 10–20% of naphtha is kerosene. Kerosene

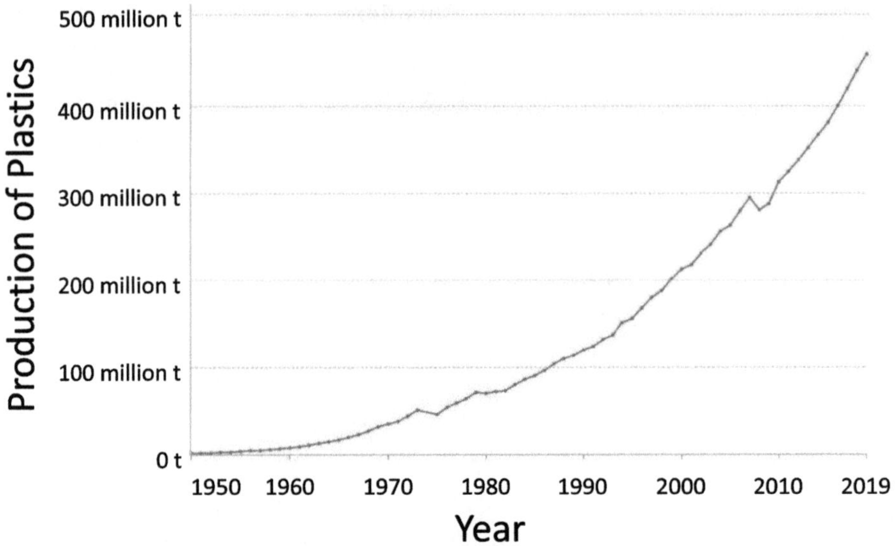

Fig. 3.7 Annual changes in plastic production. Geyer et al., Science Advance (2017) [8], OECD (2022)—with major processing by Our World in Data, adapted from https://ourworldindata.org/grapher/global-plastics-production under the Creative Commons Attribution 4.0 International License

is used as fuel for household heating and some industrial boilers. About 10–20% of naphtha is used as a raw material for petrochemicals.

3.4 Problems with Plastic Waste

Global Trends in Plastic Waste

By 2024, global plastic waste generation is projected to reach about 220 million tonnes, according to OECD projections, representing an increase of roughly 5–7% from 2021 levels. This indicates that the trend of increasing plastic waste continues (Earth Analytics). In particular, about 70 million tonnes of plastic are estimated to leak into the natural environment each year (UNEP; some estimates by BAN and Earth Analytics give similar figures depending on scope). Most of the plastic waste is sent to landfills, some are incinerated, but the recycling rate is still low. Also, at least one million tons of plastic flow into the sea every year, most of which stay near the coast, but some sink to the deep sea and accumulate as fine microplastics (Our World in Data). The amount of these plastic wastes is expected to increase further in the future, and it is said that it will increase about three times by 2060 (Basel Action Network). Strengthening the management and recycling of plastic waste is an urgent issue, and it is required that each country cooperate to take measures.

Environmental and Human Health Concerns

Discarded plastic leads to marine and land pollution. When discarded plastic flows into the ocean, marine creatures may accidentally ingest it. This can cause fish and birds to suffocate or die from blockages in their digestive tracts. Furthermore, plastic breaks down into tiny pieces (microplastics) under the influence of ultraviolet rays and waves, adversely affecting the entire ecosystem. Plastic does not decompose in the soil and remains for a long time. This leads to the pollution of farmland and natural environments, negatively impacting the ecosystem. Moreover, microplastics can potentially enter the human body through water and food. The long-term health effects are not yet fully understood, but there are concerns about their function as endocrine disruptors and their toxicity.

Challenges and Opportunities in Recycling

Most plastic is made from petroleum, and if discarded plastic is not properly recycled, petroleum resources continue to be wasted. However, the collection, sorting, recycling, and disposal of discarded plastic involves significant costs. In particular, the task of separating recyclable plastic from non-recyclable plastic is labor-intensive and increases costs. Therefore, recycling is not adequately performed. Furthermore, due to the diversity and complexity of plastic products, recycling is often difficult. In many countries, the recycling rate is low, and as a result, a lot of discarded plastic is incinerated or landfilled.

It is necessary to promote the reuse of discarded plastic by improving the sorting and recycling technology of plastic. Also, it is necessary to change the behavior of companies and consumers by strengthening regulations related to the use and disposal of plastic products. Through education and campaigns, consumers need to deepen their understanding of the problem of plastic waste and its solutions.

While CO_2 discarded into the atmosphere is certainly a problem, waste like discarded plastics thrown into the ocean or soil is also a significant issue. What they have in common is that they are disposable.

Here, when we consider a methanol society, discarded plastic can be converted into CO, CO_2, and H_2, which are raw materials for the synthesis of methanol. Plastics mainly consist of carbon, hydrogen, and oxygen atoms, and these can be gasified by reacting them with H_2O under a catalyst. The gasification of plastics into CO, CO_2, and H_2—feedstocks for methanol synthesis—is technically feasible based on existing processes, though feedstock heterogeneity and contaminant removal present operational challenges. It is just expensive. Unless we address the CO_2 problem and the discarded plastics problem at the expense of cost, the problem will not be fundamentally solved.

References

1. Hasanbeigi A (2021) Global Efficiency Intelligence: Global Steel Industry's GHG Emissions. https://www.globalefficiencyintel.com/new-blog/2021/global-steel-industrys-ghg-emissions. Accessed 11 June 2025
2. IEA (2023) P.95 Steel—breakthrough Agenda report 2023—Analysis. IEA. https://www.iea.org/reports/breakthrough-agenda-report-2023/steel. Accessed 11 June 2025
3. IEA (2023) Carbon Capture, Utilisation and Storage—Energy System—IEA. https://www.iea.org/energy-system/carbon-capture-utilisation-and-storage. Accessed 30 Dec 2024
4. The Institute for Energy Economics and Financial Analysis, IEEFA (2024) Carbon Capture for Steel? CCUS will not play a major role in steel decarbonization. https://ieefa.org/sites/default/files/2024-04/Carbon%20capture%20for%20steel-April24.pdf. Accessed 30 Dec 2024
5. IEA (2023) P.134 Cement—breakthrough Agenda report 2023—analysis. IEA. https://www.iea.org/reports/breakthrough-agenda-report-2023/cement. Accessed 11 June 2025
6. IEA (2009) Cement Technology Roadmap 2009: Carbon Emissions Reductions up to 2050. https://iea.blob.core.windows.net/assets/897b7ad9-200a-4314-b870-394dc8e6861a/CementTechnologyRoadmap-CarbonEmissionsReductionsupto2050.pdf. Accessed 30 Dec 2024
7. Bjergea LM, Brevik P (2014) CO_2 capture in the cement industry, Norcem CO_2 capture project (Norway). Energy Procedia 63:6455–6463
8. Geyer R, Jambeck JR, Lavender K (2017) Production, use, and fate of all plastics ever made law. Sci Adv 3(e1700782):1–5
9. IEA. Energy System/Industry/Chemicals. https://www.iea.org/energy-system/industry/chemicals. Accessed 30 Dec 2024
10. Gabrielli P, Rosa L, Gazzani M, Meys R, Bardow A, Mazzotti M, Sansavini G (2023) Net-zero emissions chemical industry in a world of limited resource. One Earth 6:682–704

Chapter 4
Policy Issues

Abstract This chapter examines the policy-related challenges that hinder effective responses to climate change. It begins with an overview of systemic problems in major countries, including inconsistent energy strategies and insufficient CO_2 pricing. It critiques the overreliance on electric vehicles and the unrealistic expectations surrounding a hydrogen society, pointing out the neglect of energy system integration and infrastructure needs. This chapter highlights widespread misunderstanding about energy costs and the lack of awareness regarding the scalability of renewable energy. It also addresses the tendency in academia to focus excessively on high-tech solutions while neglecting practical, scalable strategies. Finally, it emphasizes the importance of numerical literacy in policymaking and the need for interdisciplinary collaboration. This chapter argues that technological optimism must be tempered by realistic assessments of energy systems and that a methanol-centered strategy deserves more attention in global climate discourse.

Keywords Climate policy · Electric vehicles · Hydrogen society · Energy cost · Academic bias · Carbon pricing · Methanol strategy

4.1 Problems in Major Countries

Role of Major CO_2 Emitters

Figure 4.1 presents data from countries with the highest levels of CO_2 emissions. The top six emitters—China, the United States, India, Russia, Indonesia, and Brazil—collectively account for approximately 67% of global CO_2 emissions. Although these countries differ significantly in population size and GDP, it is evident that their climate policies have a decisive influence on global emission trends. CO_2 emissions are closely tied to economic development, with gross domestic product (GDP) serving as a standard indicator of economic activity. In the absence of effective carbon neutrality policies, there tends to be a proportional relationship between GDP and energy consumption, resulting in proportional CO_2 emissions.

J. Nakamura, M. Ishida, *Blueprint for a Methanol Society*,
https://doi.org/10.1007/978-981-95-3465-4_4

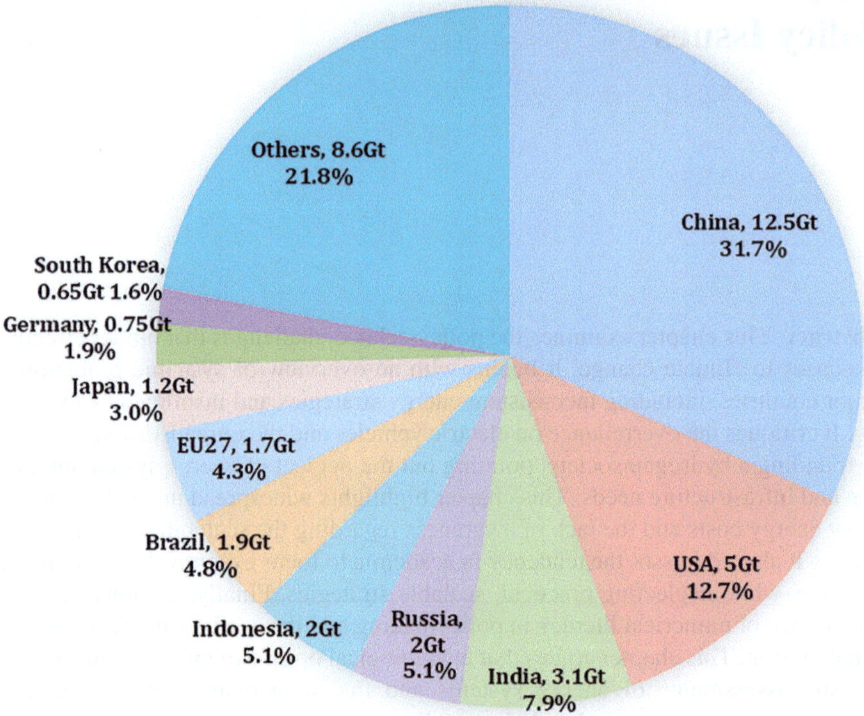

Fig. 4.1 2023 CO_2 emissions by country [1]. Data source; Global Carbon Budget (2024), Adapted from https://ourworldindata.org/co2-emissions-metrics under the Creative Commons Attribution 4.0 International License

Consequently, efforts to limit CO_2 emissions are often perceived as potential constraints on economic growth. This inherent trade-off contributes to the slow progress in implementing effective CO_2 reduction strategies. However, an examination of major emitting countries reveals that in many cases, overall, CO_2 emissions have not declined sufficiently in proportion to economic maturity, suggesting significant deficiencies or inefficiencies in their carbon neutrality policies.

Economic activity also correlates with population size, making per capita CO_2 emissions a useful indicator of emission intensity and policy effectiveness. If carbon neutrality initiatives have been effective, a notable decline in per capita CO_2 emissions should be evident in recent years. With this framework in mind, we now turn to Fig. 4.2 (per capita CO_2 emissions in several countries).

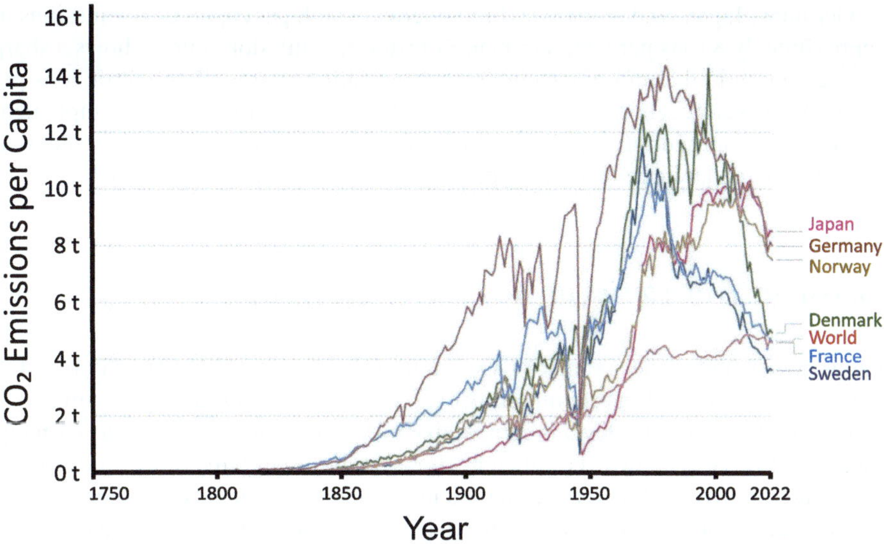

Fig. 4.2 CO_2 emission per capita in countries that reduce it in recent years [1]. Data source; Global Carbon Budget (2024), Adapted from https://ourworldindata.org/co2-emissions-metrics under the Creative Commons Attribution 4.0 International License

Regional Variations in Carbon Neutrality Efforts

First, let us consider countries where carbon neutrality progress has been relatively successful. As of 2022, the global population stands at approximately 8 billion, with the average per capita CO_2 emissions at ~5 tons per year (Fig. 4.2). Against this baseline, countries such as Sweden, Denmark, and France report per capita emissions in the range of 3.5–5.0 tons per year—down from more than 10 tons in the past—indicating substantial progress toward decarbonization. These countries benefit from regional advantages that enable the effective utilization of renewable energy sources. For example, Sweden derives the majority of its electricity from renewables, primarily hydropower and wind, while also maintaining a significant share (approximately 40%) of nuclear power. Fossil fuels, including coal, oil, and natural gas, comprise less than 20% of Sweden's primary energy supply. Denmark leads the world in wind power, which supplies over 50% of the country's electricity. Biomass contributes an additional 20%, while fossil fuels are predominantly used in the transport sector, with oil and natural gas comprising about 20% and 10% of the primary energy supply, respectively. France relies heavily on nuclear energy, which accounts for around 70% of its total electricity generation. Hydropower remains the principal renewable source at 12%, with wind energy contributing about 7% of the electricity mix.

Germany, Japan, and Norway exhibit relatively high per capita CO_2 emissions of approximately 8 tons per year. Germany's historical emissions curve shows a sharp rise beginning in the mid-nineteenth century, peaking at around 14 metric tons per capita. However, recent data indicate a steep decline, reflecting effective implementation of carbon neutrality policies. In contrast, Japan and Norway have demonstrated only moderate reductions in per capita emissions since around 2010.

Emissions Trajectories in the Largest Economies

Let us now examine the six major CO_2-emitting countries, as illustrated in Fig. 4.3. In China, the world's largest emitter, per capita CO_2 emissions are approximately 7 metric tons per year. This level is comparable to current values observed in Germany, Japan, and Norway. However, whereas those countries' figures reflect a declining trend due to active carbon reduction policies, China's emissions are still on the rise. China's carbon neutrality strategy aims to peak CO_2 emissions by 2030 and achieve net-zero emissions by 2060. In reality, however, as shown in Fig. 4.3, total annual CO_2 emissions continue to increase, and in the short term, the growth rate has accelerated following the COVID-19 recovery, although over the long term the rate of increase has been gradually slowing. Achieving emission reductions while maintaining economic development presents a formidable challenge. Given that China alone accounts for nearly one-third of global CO_2 emissions, its trajectory exerts a

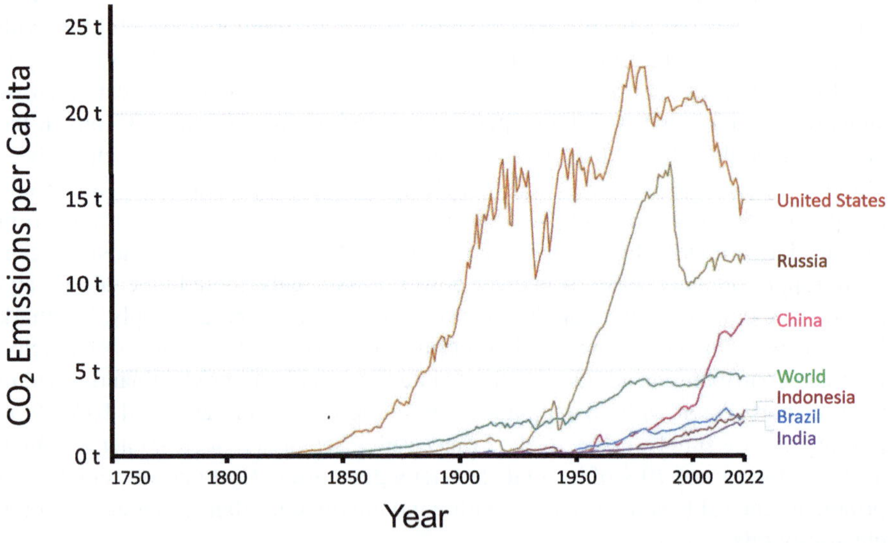

Fig. 4.3 CO_2 emission per capita in countries with increasing emission, including USA [1]. Data source; Global Carbon Budget (2024), Adapted from https://ourworldindata.org/co2-emissions-metrics under the Creative Commons Attribution 4.0 International License

profound influence on global climate efforts. China has made progress in deploying renewable energy, particularly solar and wind power, and has implemented additional measures including a national carbon emissions trading system, policies to improve energy efficiency, and research into hydrogen-based steelmaking. Despite these efforts, actual CO_2 emissions have not declined, and the country remains heavily reliant on inexpensive fossil fuels. If China fails to achieve substantial domestic reductions—despite exporting green technologies such as electric vehicles and solar panels—it risks being perceived as prioritizing economic growth over climate action. Consequently, the international community must strengthen diplomatic and policy-driven efforts to support a meaningful transition in China's climate strategy.

Turning to the United States, the second-largest emitter, there has been a notable reduction in per capita CO_2 emissions in recent years, yet the current level remains high at approximately 15 metric tons per year. The global implications of the United States' emissions trajectory are therefore substantial. U.S. climate policy focuses on expanding renewable energy capacity (solar and wind), promoting nuclear energy, adopting carbon capture and storage (CCS) technologies, and implementing a cap-and-trade emissions system. However, the effectiveness of these measures remains limited. The United States has pledged to reduce greenhouse gas emissions by 50–52% relative to 2005 levels by 2030. As shown in Fig. 4.4, U.S. CO_2 emissions in 2005 were approximately 6.35 billion tons, implying a 2030 target of roughly 3.175 billion tons. However, emissions in 2022 remain at 5.2 billion tons, indicating that current reduction efforts are insufficient to meet the target. Figure 4.4 also shows that the transportation and power generation sectors together account for nearly 60% of U.S. CO_2 emissions. While renewable energy and battery storage technologies are being deployed for grid balancing, there is a notable deficiency in

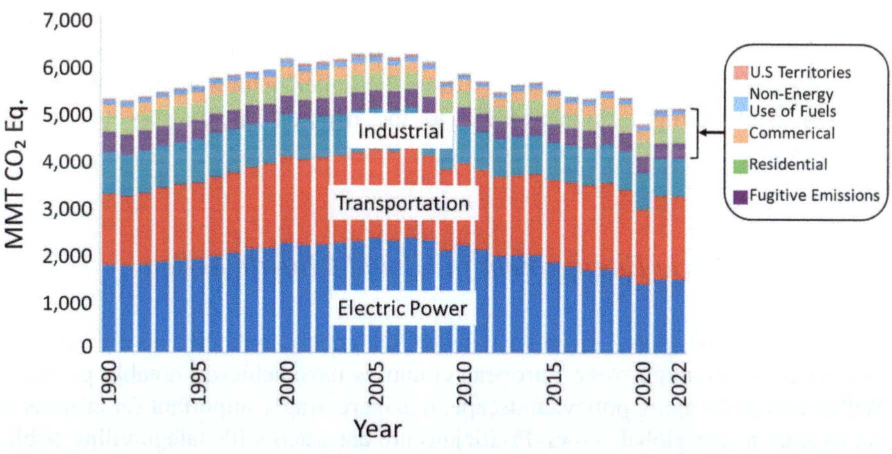

Fig. 4.4 CO_2 emission sectors in the United States [2]. Adapted from https://www.epa.gov/system/files/documents/2024-04/us-ghg-inventory-2024-main-text_04-18-2024.pdf (Fig. 2.5) under Creative Commons Zero v1.0 Universal Public Domain Dedication

the adoption of alternative gaseous and liquid energy carriers to replace fossil fuels. Although hydrogen production has been identified as a key focus area, progress has been slow and fragmented. Moreover, if future federal policies were to favor expanded fossil fuel use for energy production, this could represent a significant setback to global climate efforts.

Emissions Trends in India, Indonesia, Brazil, and Russia

Next, we examine the countries ranked third through sixth in total CO_2 emissions—namely, India, Russia, Indonesia, and Brazil. In India, Indonesia, and Brazil, per capita CO_2 emissions are approximately 2.5 metric tons per year, which is roughly half of the global average of ~5 metric tons per year. This suggests substantial potential for future economic growth. However, this potential also implies that stringent CO_2 emission controls may be perceived as constraints on economic development. The implementation of carbon neutrality measures requires substantial capital investment and policy coordination, making significant CO_2 reductions in these countries unlikely without comprehensive international support. Indeed, emissions in these countries have shown a consistent upward trend from 2012 to 2022.

Russia presents a different case. Its CO_2 emissions declined sharply following the collapse of the Soviet Union in 1988, primarily due to the economic downturn and the associated reduction in industrial activity. Since then, emissions have been increasing steadily. Unlike India, Indonesia, and Brazil, Russia currently has a high per capita CO_2 emission rate—approximately 12 metric tons per year—with no clear signs of decline. Russia's efforts toward CO_2 reduction are generally perceived as less proactive compared to other major economies. Although Russia has set a target of reducing emissions by 25–30% relative to 1990 levels by 2030, this target is based on a baseline that includes the post-Soviet industrial collapse and is aligned with an energy strategy that remains heavily dependent on fossil fuels. The deployment of renewable energy in Russia is extremely limited, with wind and solar power together contributing less than 0.1% to the national energy mix (excluding hydropower).

The Role of Civic Engagement in Global Climate Action

In conclusion, the global effort to reduce CO_2 emissions shows no encouraging signs overall, although some European countries have achieved notable progress. Within this challenging policy landscape, it is increasingly important for citizens to act as responsible global actors. Politicians are entrusted with safeguarding public welfare, and corporations are expected to contribute to societal well-being. Meaningful change toward carbon neutrality may be achieved through informed civic engagement that influences political and industrial decision-making. To enable such influence, it is essential that at least those citizens with a background in science and engineering understand the fundamental issues and quantitative data presented in this book. In the face of an escalating planetary crisis, sound decision-making depends on a solid foundation in scientific knowledge. If citizens, policymakers, and corporations collectively share an accurate and quantitative understanding and determine an appropriate course of action, real and lasting change becomes possible.

4.2 Excessive Expectations for Electric Vehicles

The deployment of electric vehicles (EVs) is frequently highlighted in national carbon neutrality policies. For example, in the United States, EV adoption is actively being promoted, although not without opposition in certain states. However, it is crucial to recognize that the introduction of EVs alone does not constitute a comprehensive carbon-neutral strategy. This section examines, considering the United States as a case study and basic calculations, why EVs cannot independently achieve significant CO_2 emission reductions.

Life-Cycle Emissions: Manufacturing and Operation Phases

To begin with, consider the CO_2 emissions associated with vehicle manufacturing. Depending on vehicle size, total emissions are estimated to be 8–12 metric tons for EVs and generally 5–10 metric tons for internal combustion engine (ICE) vehicles, although smaller or lighter ICE vehicles may emit less than 5 metric tons during manufacturing. Thus, the production of an EV generally results in approximately 20% higher CO_2 emissions. Notably, 40–60% of the CO_2 emissions from EV manufacturing are attributable to battery production. For instance, manufacturing a 50 kWh battery—typical for a mid-sized EV—can emit up to approximately 7.5 tons of CO_2, depending on the carbon intensity of the electricity used.

Next, let us compare the emissions from vehicle operation. According to major automobile manufacturers, ICE vehicles emit approximately 110–120 g of CO_2 per kilometer driven. Over a vehicle lifetime of 100,000 km, this corresponds to 11–12 tons of CO_2. Therefore, the total life-cycle emissions of an ICE vehicle—combining manufacturing and driving—fall in the range of 16–26 tons. In contrast, although EVs do not emit CO_2 directly while driving, their indirect emissions depend heavily on the electricity generation mix. In the United States, as of recent data, fossil fuel sources account for approximately 60% of total electricity production. Consequently, EV operation cannot be considered entirely emission-free.

A simple calculation of the CO_2 emissions associated with EV operation in the United States over the 10-year period from 2024 to 2034, assuming an average fossil fuel power share of 50%, is as follows:

- Power consumption per km: 0.15 kWh/km
- Total power consumption for 100,000 km: 0.15 kWh/km × 100,000 km = 15,000 kWh
- Electricity derived from fossil fuels (50%): 15,000 kWh × 0.5 = 7500 kWh
- CO_2 emission factor from fossil fuel power: 0.5 kg CO_2/kWh
- Total CO_2 emissions from EV operation: 7500 kWh × 0.5 kg CO_2/kWh = 3750 kg CO_2

Thus, even under current and near-future U.S. grid conditions, driving an EV for 100,000 km results in approximately 3.7 tons of indirect CO_2 emissions. This value must be added to the manufacturing emissions to obtain the full life-cycle CO_2 emissions of EVs.

Broader Implications for Emission Reduction Strategies

In summary, a life-cycle comparison of CO_2 emissions reveals that ICE vehicles emit approximately 16–26 metric tons, while EVs emit roughly 12–16 metric tons. Superficially, these figures suggest a potential 30–40% reduction in CO_2 emissions through EV adoption. However, this conclusion relies on the unrealistic assumption that all vehicles will be replaced by EVs within a decade. In reality, the EV adoption rate in the United States—measured as the share of battery electric vehicles (BEVs) in new vehicle sales—was approximately 7.6% between January and December 2023. If EV adoption is accelerated through mandatory replacement policies, emissions from the manufacturing phase—particularly from battery production, which alone contributes 8–12 tons of CO_2 per vehicle—will increase substantially. In many cases, continuing to operate an existing ICE vehicle may result in lower total emissions than prematurely replacing it with a new EV.

Given that the next 10–20 years are critical for climate change mitigation, strategies must prioritize rapid and substantial CO_2 reductions. However, current carbon neutrality policies that rely heavily on EV deployment fall short of meeting this objective. The United States, for instance, emits approximately 5 billion metric tons of CO_2 annually, with the transportation sector contributing about 30%—equivalent to 1.5 billion tons. Even under the optimistic assumption of a 20% reduction in transportation emissions, the net impact would amount to a decrease of only 300 million tons—a modest reduction for a leading industrialized nation.

This limitation is not unique to the United States. Globally, there is a widespread perception that EVs and batteries are central to the transition to renewable energy. However, this perspective overlooks several critical challenges. First and foremost is the issue of resource availability. If the global vehicle fleet were to be electrified using lithium-ion batteries, demand for key materials such as lithium (Li), cobalt (Co), and nickel (Ni) could approach or exceed current known reserves for Li and Co over the coming decades.

Moreover, automotive batteries typically weigh between 350 and 700 kg for most passenger vehicles, though some large electric SUVs and trucks may use battery packs approaching 900 kg. This is about two to three times heavier than a conventional ICE system (including engine and fuel tank, typically 100–300 kg). For a medium-sized passenger car, an EV is typically about 200–300 kg heavier than an ICE vehicle, with some large models exceeding this by a greater margin. This weight increase reduces overall energy efficiency. In addition, there are significant challenges associated with the high cost of developing charging infrastructure, the limitations of current battery recycling technology, and the looming depletion of critical mineral resources.

While research is ongoing to develop alternative materials to Li and Co, no viable replacements have emerged to date. The fundamental problem remains that EV production demands substantial material resources. Furthermore, electricity supply capacity from existing power grids is already constrained. In some cases—such as in Switzerland and certain regions of the United States—grid limitations have already hindered EV charging availability.

Given these challenges, a complete transition to EVs appears neither feasible nor sustainable in the near future.

4.3 Lack of Understanding of a Hydrogen Society

The integration of renewable energy is essential for achieving carbon neutrality. A key challenge, however, lies in determining whether the generated electricity should be stored in the form of batteries or as chemical substances. Among the latter, hydrogen has attracted significant attention as a potential energy carrier. While this book proposes methanol as an alternative, the concept of hydrogen carriers and methanol carriers is closely related. The central argument here, however, is that relying solely on hydrogen as an energy medium presents significant limitations.

At first glance, carbon neutrality through hydrogen utilization may evoke the image of a clean, sustainable society. Nevertheless, this section aims to demonstrate—through quantitative analysis—just how challenging it is to realize a so-called hydrogen society. In 2022, the President of the United States declared the ambition to make the country a "hydrogen superpower," setting a goal of producing ten million tons of hydrogen annually by 2030, with an investment of approximately 1 trillion yen dedicated to hydrogen production alone.

Quantitative Analysis of Hydrogen's CO_2 Reduction Potential

To estimate the CO_2 reduction potential of this hydrogen production, we consider the enthalpy change associated with combustion. Oil and natural gas generally consist of hydrocarbons with the formula n CH_x ($x = 2$–4) and their complete oxidation releases on average approximately 800 kJ of energy per mole of CO_2 emitted. In contrast, the combustion of hydrogen $\left(H_2 \frac{1}{2} O_2 \rightarrow H_2O\right)$ releases 285.8 kJ/mol.

Therefore, to generate the same amount of energy per mole of CO_2 emission (44 g), approximately 5.66 g of hydrogen is required.

Given that annual CO_2 emissions are approximately 5 billion tons in the United States and 1.1 billion tons in Japan, the amount of hydrogen required to offset these emissions would be about 640 million tons (United States) and 140 million tons (Japan), respectively. Thus, even if the United States achieves its ten million ton hydrogen production target by 2030, this would offset only about 1.6% of national CO_2 emissions. Although this is a rough estimate, the result is so modest that further adjustment with more detailed parameters would not significantly change the conclusion. In short, an enormous quantity of hydrogen is required to meaningfully reduce CO_2 emissions.

Anyone with a basic understanding of chemistry should recognize this limitation. When hydrocarbons are combusted, both CO_2 and energy are produced. Comparing this with the energy obtained from hydrogen combustion, it becomes apparent that, despite hydrogen's low molecular weight (2 g/mol) relative to CO_2 (44 g/mol), achieving an equivalent energy output would require several billion tons of hydrogen annually to match the combustion heat associated with the world's annual CO_2 emissions of approximately 40 billion tons. Therefore, producing only tens of millions of tons of hydrogen is far from sufficient.

The Need for Rigorous Quantitative Evaluation

The authors do not entirely reject the use of hydrogen carriers. However, we argue that certain forms of hydrogen utilization—such as liquefied hydrogen—are impractical and lack justification unless accompanied by rigorous quantitative evaluation. Without such scientific analysis, hydrogen policy risks becoming ineffective and failing to contribute to the goal of carbon neutrality.

4.4 Lack of Energy Cost Awareness

A comprehensive understanding of energy efficiency is essential when evaluating carbon neutrality strategies. However, basic energy analyses are often absent from public policy discussions. This lack of visibility represents a significant issue. One illustrative case is the coal gasification process for hydrogen production, which has been considered in Japan in recent years. This process involves gasifying coal and reacting it with water vapor to produce hydrogen, with the resulting CO_2 separated and stored underground via carbon capture and storage (CCS). This section examines the inefficiencies and resource losses inherent in this approach. The conclusion is that, as shown in Fig. 4.5, significant energy losses occur in the process of producing hydrogen from coal. This is discussed in more detail below.

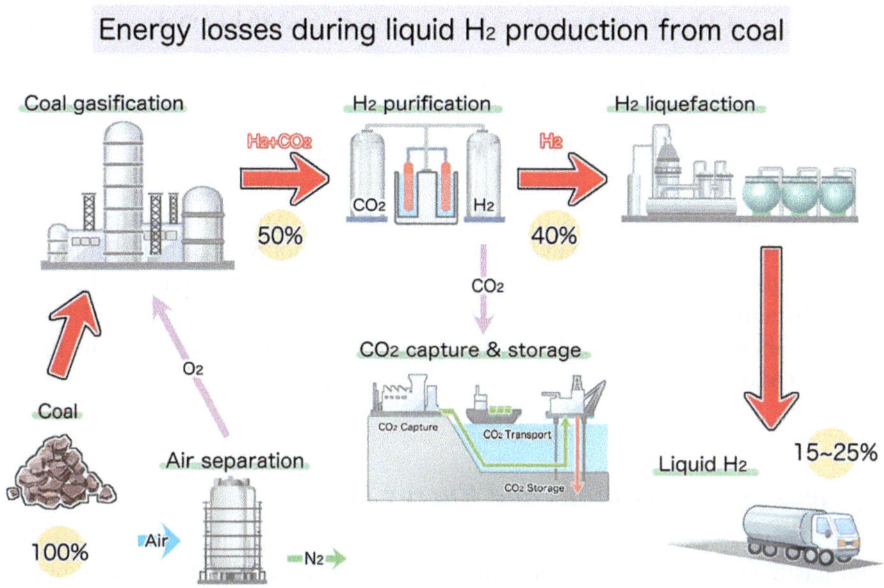

Fig. 4.5 Energy losses during liquid H_2 production from coal

Thermodynamic Basis of Coal Gasification

Coal gasification is a high-temperature process (typically above 700 °C) in which pulverized coal reacts with oxygen, air, or steam to generate syngas—a mixture primarily composed of hydrogen (H_2) and carbon monoxide (CO). The principal reaction is:

$$C + H_2O(l) \rightleftharpoons H_2 + CO \qquad \Delta_r H^\circ = +175.3\,\text{kJ/mol} \tag{4.1}$$

This is an endothermic reaction and thus requires heat, typically supplied by partial combustion of coal:

$$C + O_2 \rightleftharpoons CO_2 \qquad \Delta_r H^\circ = -393.5\,\text{kJ/mol} \tag{4.2}$$

Combining these, the approximate overall reaction for syngas production with heat balance becomes:

$$1.445C + 0.445O_2 + H_2O(l) \rightleftharpoons H_2 + CO + 0.445CO_2 \qquad \Delta_r H^\circ \approx 0 \tag{4.3}$$

In practice, however, more carbon and oxygen are required to account for sensible and latent heat of reactants and heat losses from the reactor. The CO produced is typically converted to H_2 via the water-gas shift reaction:

$$CO + H_2O(l) \rightleftharpoons H_2 + CO_2 \qquad \Delta_r H^\circ = 2.86\,\text{kJ/mol} \tag{4.4}$$

When heat-balanced ideally, the net equation becomes:

$$1.452C + 0.452O_2 + 2H_2O(l) \rightleftharpoons 2H_2 + 1.452CO_2 \qquad \Delta_r H^\circ \approx 0 \tag{4.5}$$

This stoichiometry implies that producing 4 g of hydrogen requires approximately 17.4 g of carbon. However, real-world coal is not pure carbon. Bituminous coal, though high in carbon content, is more energy-dense than lignite, which contains substantial moisture and other impurities. The lower heating value (LHV) of carbon is about 32.8 MJ/kg, whereas bituminous coal yields 27.8 MJ/kg and lignite only 12.0 MJ/kg.

Efficiency Considerations and Energy Losses

Coal gasification can improve thermal power plant efficiency through combined cycle generation using gas and steam turbines. However, if syngas is extracted for hydrogen production instead of being used directly for power generation, substantial energy is lost—particularly due to gas cooling and hydrogen separation processes. Hydrogen must be separated at near ambient temperatures, which requires significant cooling energy.

Reported conversion efficiencies from coal to hydrogen are approximately 40–50% [3]. For instance, the specific coal consumption per ton of hydrogen is 21.0 t/t-H_2 for lignite and 8.7 t/t-H_2 for bituminous coal. Assuming the energy contents mentioned above, these correspond to stoichiometric efficiencies of 47.6% and 49.6%, respectively.

Further losses arise during CO_2 separation in the CCS process. Even under idealized stoichiometric conditions (Eq. 4.5), the product stream following the water-gas shift reaction is estimated to contain approximately 42.1% CO_2 by volume. However, the actual CO_2 concentration depends on the specific gasification process, operating conditions, and extent of conversion and may vary significantly across different system configurations. Capturing and compressing this CO_2 introduces an "energy penalty," typically estimated at 10–30%, depending on the process design, technical specifications, and operational conditions. A general approximation is that about 20% of total energy is lost to CCS.

Hydrogen Liquefaction and Final Energy Retention

Hydrogen liquefaction imposes additional energy losses. Conventional cryogenic processes—such as the Claude and Linde cycles—require multiple stages of compression, heat exchange, and expansion. Despite decades of development, these processes offer limited room for efficiency improvement. Even with idealized assumptions, liquefaction typically consumes 30–60% of the hydrogen's energy content (based on primary energy).

When evaluating hydrogen production from coal with 50% conversion efficiency, subtracting ~20% for CCS and ~ 30 to 60% for liquefaction leaves only about 15–25% of the original coal energy in usable liquefied hydrogen. How one interprets Fig. 4.5 is crucial in guiding future technological and policy directions.

Practical Limitations and Misconceptions

According to a literature [3], even under optimistic future scenarios with advanced CCS achieving 88% CO_2 capture, the overall emissions would still equate to approximately 70% of the emissions from directly burning natural gas without CCS, based on life-cycle assessments that include residual emissions, upstream processes, and the energy penalty of capture and compression. This means that hydrogen from coal gasification would reduce CO_2 emissions by only 30% compared to unabated natural gas combustion—a figure far from the ideal of "carbon-free hydrogen."

Achieving meaningful carbon neutrality through coal-based hydrogen production would require CCS with capture rates exceeding 99%. This underscores the technological and economic fragility of relying on energy-inefficient processes like coal gasification and CCS as the foundation of hydrogen energy systems.

4.5 Issues in Academia

Although universities and research institutions have contributed significantly to fundamental research, there remains a shortage of scalable, economically viable solutions that can be rapidly deployed for deep decarbonization. Numerous technical advancements have been made in areas such as photovoltaics and electrochemical systems, but few have reached the level of economic and industrial maturity required for large-scale deployment. While society expects academia to deliver breakthrough innovations to combat climate change, the reality is that few tangible solutions have emerged in recent decades that address emissions reduction at the required scale. This highlights a structural issue within the academic sector.

One underlying reason is the intrinsic nature of academic research, which often emphasizes novelty and originality over applicability. Researchers are driven by the pursuit of unique and academically interesting topics. While this is not inherently negative, it is clear that such research alone cannot solve the existential crisis of climate change. Much of the work in universities is structured around publishing papers, rather than solving real-world energy or environmental challenges. For example, in materials research related to energy carriers, much focus is placed on the development of high-performance batteries, advanced catalysts, and separation membranes. However, many of these studies rely on costly or scarce materials, such as platinum or iridium—priced at over US$10 per gram—which renders them economically unviable for large-scale implementation. Over the past two decades, substantial research funding has been allocated to such material science investigations, yet no transformative innovation has emerged. While such research may become useful in the future, it often fails to address the immediate needs of global warming mitigation, especially in terms of economic feasibility and deployment at scale.

There is a growing need for scientists and engineers capable of engaging with real-world problems—including cost, energy systems, and policy implications—rather than remaining focused solely on theoretical or elemental research. High-tech solutions that are only applicable in developed nations are no longer sufficient. A new philosophy is needed—one that defines innovation as technology capable of delivering global impact. This calls for researchers who prioritize problem-solving over academic prestige and a shift in science and technology policy to support such individuals. To this end, large-scale investment must be directed toward identifying and nurturing talent committed to solving urgent societal challenges. Policies should encourage broad, quantitative, and scientifically grounded approaches to achieve genuine progress in carbon neutrality.

Structural Limitations of Academic Research Systems

The prevailing academic evaluation systems—focused primarily on publication quantity and citation-based metrics—have led to a culture where novelty and specialization are prioritized over practical impact. As a result, the current publication-driven evaluation system often disincentivizes interdisciplinary, systems-level research essential for addressing complex challenges like climate change. This narrow focus has contributed to an overemphasis on materials science while neglecting the broader system integration necessary for real-world energy transitions.

Materials science is undoubtedly critical for developing new energy materials, high-efficiency conversion systems, and energy storage technologies. It plays a central role in improving the performance and reducing the cost of technologies such as solar power, fuel cells, batteries, and carbon capture and storage (CCS). However, materials innovation alone cannot create sustainable energy systems or a decarbonized society. A systems engineering approach is equally essential and must be integrated into R&D efforts.

Key components of a systems approach include:

i. **Integrated System Design:** Optimizing energy generation, storage, distribution, and use as a unified whole.
ii. **Infrastructure Compatibility:** Ensuring that new energy technologies are interoperable with existing infrastructure.

iii. **Policy and Regulation Frameworks:** Establishing supportive regulatory environments for emerging technologies.
iv. **Market and Social Acceptance:** Addressing user acceptance and market integration for new technologies.
v. **Value Chain Perspective:** Considering the entire life cycle from development to deployment to enhance sustainability and cost-effectiveness.

Achieving carbon neutrality demands the coordinated advancement of both materials science and system-level design. Policymakers, researchers, and industry stakeholders must foster an environment where these fields are developed synergistically, not in isolation.

Reforming Research Institutions and Evaluation Systems

Academic institutions are increasingly constrained by an overemphasis on elemental research. There is insufficient support for holistic, systems-based R&D, due in part to structural and institutional barriers. The current evaluation system rewards short-term, publication-centric achievements and discourages long-term, interdisciplinary projects. System-level research typically requires longer timelines, more funding, and broader collaboration—all of which make evaluation and funding more complex. Additionally, collaborative and interdisciplinary research remains difficult due to institutional cultures that often do not reward such efforts. In this environment, complex systems research struggles to gain traction.

To address these issues, the academic evaluation framework must be revised to recognize and reward long-term, impactful research—especially efforts that contribute to societal and industrial transformation. Government and private funding should be expanded to support not only basic science but also applied and systems-level research. Interdisciplinary and cross-sector collaboration mechanisms must be strengthened, both within institutions and across academia, industry, and government.

In summary, to build a sustainable research and development ecosystem that effectively contributes to science, technology, and society, all stakeholders—policymakers, research institutions, and industry—must work together to realign incentives, restructure evaluation systems, and prioritize impact-oriented research.

4.6 Excessive Expectations for Science and Technology

While the advancement of science and technology has brought remarkable benefits to humanity—ranging from medical breakthroughs to improved food security and enhanced access to information—it is essential to acknowledge that science and technology are not all-powerful solutions to every human challenge. Their potential is bounded by physical, ethical, resource, social, and psychological limitations.

Physical Limitations

Scientific and technological progress must adhere to the immutable laws of nature. For instance, with our current understanding, no mechanism allows for communication or transportation faster than the speed of light. Fundamental physical laws constrain what is scientifically achievable, regardless of technological ambition.

Ethical Limitations

Technological solutions often raise complex ethical questions. For example, while gene-editing technologies such as CRISPR offer promising avenues for treating diseases, their use prompts critical ethical considerations, including genetic equity, long-term effects, and societal acceptance.

Resource Limitations

Earth's natural resources are finite. No matter how advanced our technology becomes, it cannot create infinite resources. Sustainable development must be pursued in parallel with scientific innovation to ensure long-term environmental and economic stability.

Social and Psychological Limitations

Human happiness and fulfillment cannot be measured solely through material or technological progress. Aspects of well-being—such as mental health, social connections, and a sense of purpose—are not addressed through science and technology alone.

Technological Potential and Its Limits

Technological innovation has undoubtedly driven efficiency improvements in many areas. However, there is growing consensus that conventional energy conversion technologies, such as fossil fuel combustion and photovoltaic systems, are approaching thermodynamic and economic efficiency limits. While emerging technologies—such as perovskite-based solar cells, quantum dots, quantum computing, and nuclear fusion—may offer paradigm-shifting potential in the long term, they remain largely in developmental stages.

In contrast, the urgency of climate change requires short-term, practical solutions. While it is important to invest in visionary technologies, immediate actions based on existing technologies and strategies are essential. These include forest conservation and afforestation, the implementation of carbon pricing mechanisms, regulatory policies, and behavioral changes at the individual and societal levels.

The Risks of Technological Fix Thinking

Overreliance on technological solutions—a mindset known as the "technological fix"—can be problematic. The belief that future innovations will solve current problems may lead to complacency and resistance to necessary behavioral and structural changes.

Key concerns include:

- **Delayed Behavioral Change**: Confidence in future technologies may lead individuals and policymakers to postpone necessary emissions reductions and resource conservation, underestimating the urgency of action.
- **Uncertainty in Outcomes**: All technological development involves inherent uncertainty. Technologies that appear feasible in theory may face insurmountable barriers in practice—technical, economic, or societal—that prevent them from achieving the intended impact.
- **Overlooked Environmental Impacts**: The production and deployment of new technologies may introduce new environmental and social burdens. For example, while electric vehicles reduce fossil fuel use, the extraction of rare earth metals for battery production can cause environmental degradation and human rights issues.

A Balanced and Realistic Approach

Effective climate action must strike a balance between short-term pragmatism and long-term vision. The scale of global energy consumption demands coordinated, bold action across nations. Although immediate global transformation may not be realistic, incremental efforts—such as changes in local behavior and regional policy—can collectively generate substantial impact over time through a "trickle-down" effect.

Historical reliance on fossil fuels has underpinned industrial development for more than a century. The prioritization of economic growth and short-term profits has reinforced dependence on coal, oil, and natural gas. Reversing this trend requires not only technological innovation but also a shift in values, policy, and societal behavior.

References

1. Global Carbon Budget (2024) Population based on various sources (2024)—with major processing by Our World in Data. https://ourworldindata.org/co2-emissions-metrics. Accessed 11 June 2025
2. EPA (2024) Inventory of U.S. Greenhouse Gas Emissions and Sinks: 1990–2022. U.S. Environmental Protection Agency, EPA 430-R-24-004. Fig. 2.5. https://www.epa.gov/system/files/documents/2024-04/us-ghg-inventory-2024-main-text_04-18-2024.pdf Accessed 3 September 2025. https://www.epa.gov/ghgemissions/inventory-us-greenhouse-gas-emissions-and-sinks-1990-2022
3. Strategy for Technology Development, Proposal Paper for Policy Making and Governmental Action toward Low Carbon Societies, Center for Low Carbon Society Strategy, Japan Science and Technology Agency (2019) Economy of Hydrogen and Ammonia by Coal Gasification and CO_2 Emissions: Study on Production and Logistics of Hydrogen and Ammonia by Coal Gasification with CCS

Part II
Which Way Should We Go?: A Methanol Society

Chapter 5
Why a Methanol Society?

Abstract This chapter explores methanol as a versatile fuel that bridges the gap between fossil fuel dependence and a carbon-neutral society. Methanol can be used directly as a fuel in internal combustion engines, blended with gasoline, or employed in fuel cells. Compared to hydrogen and ammonia, methanol offers advantages in storage, transportation, and infrastructure compatibility due to its liquid state under ambient conditions. This chapter examines energy density, combustion characteristics, and emission profiles, including reduced particulate matter and NO_x emissions. It also reviews various methanol-fueled applications, from vehicles to marine engines and boilers. Methanol is already in commercial use in some regions, especially China, and shows strong potential for broader adoption. Crucially, in a methanol society, renewable energy can be used to synthesize methanol from captured CO_2, creating a closed-loop system where carbon resources are continually recycled. This approach enables not only decarbonized fuel use but also a sustainable and circular carbon economy. This chapter emphasizes that methanol-based energy systems are both technically feasible and environmentally advantageous, providing a realistic path toward global decarbonization.

Keywords Methanol fuel · Carbon recycling · Energy density · Internal combustion · Fuel cells · Infrastructure · Circular economy

5.1 CO_2–Methanol Cycle Society

In light of the accelerating reality of global warming, the urgent reduction of CO_2 emissions has become a critical global challenge. As discussed previously, the majority of CO_2 emissions originate from energy-related activities. Therefore, the first priority should be the prompt capture of CO_2 at its emission sources. During the transition from fossil-based energy systems to those based on renewable sources, incorporating the recovered CO_2 into an energy carrier can significantly contribute to emission reduction. In Part I, we emphasized the importance of converting electricity generated from renewable energy into molecular forms for storage and

J. Nakamura, M. Ishida, *Blueprint for a Methanol Society*,
https://doi.org/10.1007/978-981-95-3465-4_5

transport and examined the role of hydrogen as a potential energy carrier. However, hydrogen presents serious logistical challenges, particularly regarding its storage and transportation. While liquefied hydrogen has been proposed to address volume-related issues associated with gaseous hydrogen, it must be maintained at an extremely low temperature (−252.9 °C), which requires substantial energy for liquefaction. This results in a considerable reduction in overall energy efficiency. As previously introduced, several hydrogen-based energy carrier candidates have been proposed, including methylcyclohexane (MCH), methane, ammonia, and methanol. Among these, methanol emerges as the most promising option. The reasons for this will be detailed below.

The Methanol-Based Circular Economy Concept

An overview of the envisioned methanol-based society is illustrated in Fig. 5.1. A key feature of this system is its dual capability to transport hydrogen and recycle CO_2. As discussed earlier, hydrogen can be produced using renewable energy sources. This hydrogen, when combined with CO_2, can be efficiently converted into methanol via the following reaction:

$$CO_2 + 3H_2 \rightleftharpoons CH_3OH(g) + H_2O(g) \quad \Delta_r H^\circ = -49.5\,\text{kJ/mol} \tag{5.1}$$

$$CO_2 + 3H_2 \rightleftharpoons CH_3OH(l) + H_2O(l) \quad \Delta_r H^\circ = -131.0\,\text{kJ/mol} \tag{5.2}$$

Water is generated as a by-product; however, methanol can be transported either in its pure form or as an aqueous solution, depending on application needs. Methanol

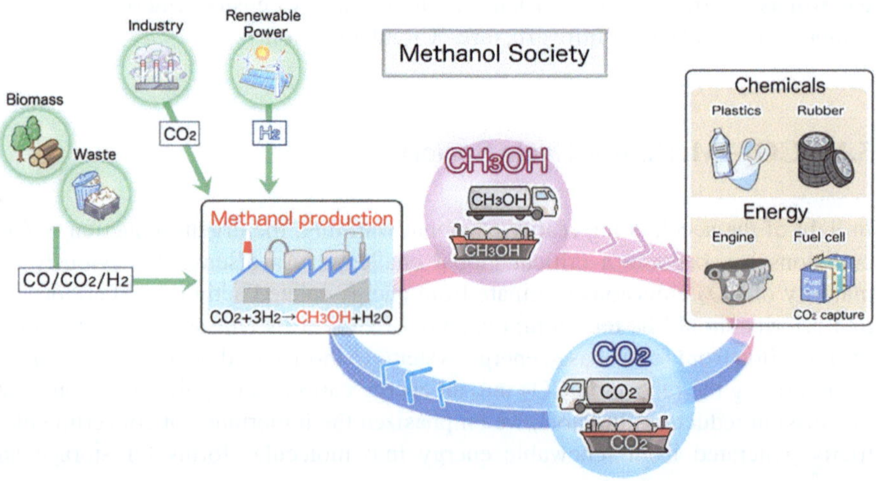

Fig. 5.1 CO_2–methanol recycling society

can be used to generate electricity via fuel cells or internal combustion engines. In fuel cell applications, solid oxide fuel cells (SOFCs) and proton exchange membrane fuel cells (PEFCs) are commonly employed. Methanol may be utilized directly as a fuel, or it can be reformed into CO_2 and H_2, with the latter used in the fuel cell.

The CO_2 generated during methanol utilization (e.g., in fuel cells) can be easily captured at the exhaust outlet, then transported—via pipeline or tanker—to methanol synthesis facilities for reuse. This enables the establishment of a closed-loop CO_2–methanol cycle.

Furthermore, methanol has potential as a high-octane, clean-burning fuel that does not emit nitrogen oxides (NO_x) or sulfur oxides (SO_x) when used in internal combustion engines. In addition, dimethyl ether (CH_3OCH_3), which can be readily synthesized from methanol, is also recognized as an excellent automotive fuel.

The infrastructure and technologies required for methanol storage and transportation are already well established, and methanol itself poses a lower safety risk compared to many other chemical fuels. These attributes make methanol a highly practical and scalable energy carrier within the framework of a carbon-neutral society.

Integration with Natural and Industrial Cycles

The widespread implementation of a methanol-based system would enable the circulation and effective utilization of a significantly larger volume of CO_2. In nature, CO_2 is continuously cycled through various biological and geochemical processes. The concept of a methanol economy closely parallels this natural CO_2 cycle. In recent years, global efforts have been initiated to integrate methanol into carbon-neutral strategies; however, these efforts remain limited in scale. There is an urgent need to expand the deployment of methanol-related technologies on a global scale.

A key characteristic of the methanol society shown in Fig. 5.1 is the establishment of a circular economy. This circulation can be classified into two major phases: the **energy cycle** and the **material (carbon) cycle**.

- In the **energy cycle**, CO_2 and methanol are recycled in tandem with the consumption and conversion of renewable electricity. Hydrogen produced via electrolysis using renewable electricity reacts with captured CO_2 to synthesize methanol. This methanol is subsequently used as a fuel in fuel cells or combustion engines. The CO_2 emitted during energy use is then recaptured and combined with hydrogen once again to produce methanol, thus forming a closed-loop energy system.
- In the **material cycle**, methanol functions as a versatile chemical feedstock. In the chemical industry, methanol synthesized from CO_2 can be converted into a range of value-added chemicals such as ethylene, propylene, and formaldehyde. These intermediates are widely used in the production of plastics and other pet-

rochemical products. After their use, waste plastics can be incinerated, and the emitted CO_2, CO, and H_2 can be recovered and used to re-synthesize methanol. This approach facilitates a carbon loop within the chemical industry, wherein CO_2 is continuously recycled in both the energy and material domains.

This circulation-based approach aligns with the carbon cycle in natural ecosystems, where biotic and abiotic processes maintain equilibrium. The anthropogenic CO_2 problem arises from disruptions to this cycle. A methanol society aims to restore and mimic the natural carbon balance, and it is increasingly recognized as a rational direction for future human civilization.

Technological Feasibility

From a technological perspective, the transition to a methanol-based society does not pose significant challenges. The core technologies required for methanol production and utilization are already well established, and in many respects, the system can be considered low-tech. Methanol synthesis is typically carried out using Cu/ZnO-based catalysts, and the conversion of CO_2 to methanol proceeds efficiently at temperatures above 500 K.

Methanol (CH_3OH), composed of carbon, hydrogen, and oxygen, is a highly versatile compound. It can be used as a clean fuel for both fuel cells and internal combustion engines and also serves as a feedstock for synthesizing petrochemical products such as ethylene, propylene, and formaldehyde. In this respect, methanol exhibits functionality comparable to that of crude oil, while offering the added advantage of carbon recyclability.

Furthermore, since the methanol production and recycling processes are powered by renewable energy, the methanol society aligns well with the principles of sustainability and carbon neutrality. Thus, the development of a methanol-based circular economy represents a scientifically sound and technologically feasible pathway toward a sustainable future.

Historical Perspective

The concept of utilizing methanol to mitigate CO_2 emissions is not new. As early as the 1990s, research efforts into methanol synthesis from CO_2 were already active. One of the authors of this book, Junji Nakamura, has been engaged in catalyst research for CO_2-to-methanol conversion since 1993, in collaboration with Dr. Tadahiro Fujitani under the RITE project at the former Research Institute for Resources and Environment. Between 1994 and 2004, a total of 34 scientific papers on methanol synthesis were published, and foundational studies on this topic have continued since that time [1–8].

To the best of the authors' knowledge, Japan was among the first countries to initiate a national-level project aimed at developing methanol synthesis from CO_2 as a climate change mitigation strategy. In 2006, Nobel laureate Dr. George A. Olah published *Beyond Oil and Gas: The Methanol Economy*, strongly advocating for a future society centered on methanol [9–13]. This vision was further supported by Dr. John O'M. Bockris, a leading electrochemist, who contributed scholarly works on the methanol economy well before 2010 [14–16].

Although several decades have passed since these pioneering contributions, efforts to utilize methanol have persisted and evolved. Developments in science and technology, shifts in public policy, and changes in global energy perspectives have collectively shaped the current landscape. Against the backdrop of accelerating global warming, it is now imperative to assess how far preparations for a methanol-based society have progressed and to explore viable paths forward.

Key Advantages and Future Outlook

The primary aim of this book is to provide a comprehensive assessment of the present status of methanol-related technologies and to explore the feasibility of realizing a methanol society within the next 20 years. While the foundational concept aligns with the visions proposed by Dr. Olah and Dr. Bockris, this work adds a new dimension by quantitatively evaluating the practicality, performance, and comparative advantages of methanol over other carbon-neutral technologies. Key metrics include energy volumetric density, ease of storage and transport, life-cycle CO_2 emissions, and compatibility with current infrastructure.

For example, volumetric energy density—the amount of energy transported per tanker—is a crucial criterion for evaluating energy carriers. Other topics examined include the stability of power supply and demand, long-term global scalability, and the establishment of robust supply chains. Particular attention is given to the use of methanol in solid oxide fuel cells (SOFCs) and internal combustion engines. Compared to polymer electrolyte fuel cells (PEFCs), which are currently more common, SOFCs offer higher power generation capacity. By combining SOFCs and PEFCs, a stable and flexible power supply can be achieved. Furthermore, the CO_2 emitted from SOFCs can be readily separated, captured, and recycled for methanol synthesis.

As discussed in Sect. 2.3, one of the principal advantages of methanol over other hydrogen carriers lies in its ease of handling under ambient conditions and its relatively high volumetric energy density. These attributes make methanol particularly well-suited for practical implementation.

Advantages of a Methanol Society

The merits of a methanol-based energy and materials system can be summarized as follows:

i. Immediate CO_2 mitigation through the establishment of a recycling system
ii. Commercial readiness, enabling prompt deployment
iii. Established infrastructure for storage and transport
iv. Proven synthesis technology, requiring minimal adjustments
v. Low toxicity and favorable energy density compared to alternatives
vi. Versatile industrial applications, fostering independence from fossil fuels
vii. Global scalability, enabling deployment even in developing countries

5.2 What Is Methanol?

Physical and Chemical Properties (Table 5.1)

Methanol (CH_3OH), also known as methyl alcohol, is a colorless, highly volatile liquid with a molecular weight of 32.04 g/mol. It has a boiling point of 64.7 °C and a melting point of −97.6 °C, with a density of 0.791 g/cm^3 at 20 °C. Its vapor pressure is 13.02 kPa (20 °C), and its viscosity is 0.545 mPa·s (20 °C). Methanol is highly soluble in water and many organic solvents, including acetone, benzene, and chloroform.

Chemically, methanol is easily oxidized to produce formaldehyde and formic acid. It is widely used as a solvent, antifreeze, and fuel and is commonly employed

Table 5.1 Properties of methanol

Property	Value
Molecular formula	CH_3OH
Molecular weight	32.04 g/mol
Melting point	−97.6 °C
Boiling point	64.7 °C
Density	0.791 g/cm^3 at 20 °C
Vapor pressure	13.02 kPa at 20 °C
Viscosity	0.545 mPa·s at 20 °C
Solubility	Soluble in water, acetone, benzene, chloroform
Toxicity	Toxic if ingested, can cause optic nerve damage
Flammability	Highly flammable, produces combustible gas

as a fuel for racing vehicles, as well as a precursor in chemical synthesis. It also serves as fuel for camping stoves.

Figure 5.2 illustrates a combustion experiment using a 1:1 molar mixture of methanol and water. This mixture burns quietly and is non-explosive in liquid form, indicating that methanol–water solutions can be relatively safe to handle under controlled conditions. However, they remain flammable, and their vapors can form explosive mixtures with air within a concentration range of 6.0–36.5% (by volume). Such mixtures may ignite or explode in the presence of a spark or heat source, requiring strict handling procedures and appropriate safety precautions.

While it is less acutely irritating than ammonia, methanol is highly toxic if ingested, inhaled, or absorbed through the skin. Once in the body, methanol is metabolized in the liver to formaldehyde and then to formic acid, which can cause metabolic acidosis and damage to the optic nerve, potentially resulting in permanent blindness. The estimated lethal oral dose in humans is 0.3–1 g/kg body weight, corresponding to approximately 30–100 mL for an adult. Symptoms of acute methanol poisoning include headache, dizziness, nausea, visual disturbances, and, in severe cases, coma and death. Chronic exposure to low concentrations may also lead to neurological and visual impairments. Occupational exposure limits set by agencies such as OSHA and ACGIH are typically 200 ppm (260 mg/m^3) as an 8-hour time-weighted average. Although methanol is widely used as a fuel and chemical feedstock, strict handling procedures, proper ventilation, and protective equipment are essential to ensure safety in production, storage, and use.

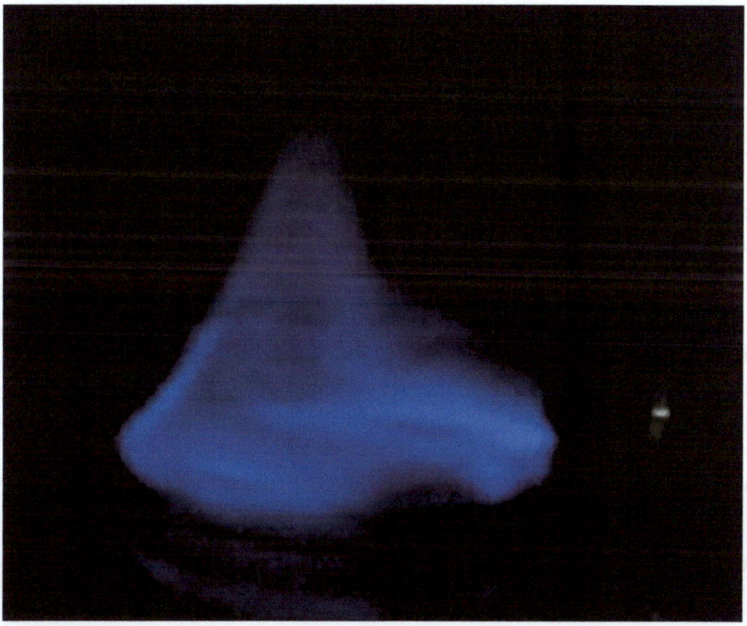

Fig. 5.2 Non-explosive methanol–water combustion

From an environmental standpoint, methanol is biodegradable, but large-scale releases can cause localized environmental pollution.

Global Production and Synthesis Methods

As of 2024, global methanol production is approximately 110 million tons per year, with China accounting for around 60–70% of total output. Methanol is produced primarily through two industrial processes:

Natural gas-based synthesis: This is the most widespread method globally. Natural gas (mainly methane) undergoes steam reforming to generate synthesis gas—a mixture of CO and H_2—via the reaction: $CH_4 + H_2O \rightleftharpoons CO + 3H_2$. The synthesis gas is then converted to methanol using a metal catalyst. This method is favored due to its high energy efficiency and low production cost.

Coal-based synthesis: In coal-rich regions, especially in China, coal is gasified to produce synthesis gas, which is subsequently converted to methanol using a similar catalytic process. While this approach is cost-effective in coal-abundant areas, it has a significantly higher environmental impact due to greater CO_2 emissions. For this reason, coal-based methanol production is restricted or banned in countries like the United States.

In parallel, research and development are progressing on renewable methanol production from biomass or captured CO_2, but these technologies remain at a pilot or small-scale stage. Nevertheless, renewable methanol is expected to play an increasingly important role as a sustainable energy carrier in the future.

Applications of Methanol

Methanol's chemical versatility makes it essential in a wide range of industrial applications.

- The largest use of methanol is as a feedstock for formaldehyde, which accounts for approximately 22% of global methanol consumption (see Fig. 5.3). Formaldehyde is a precursor for plywood, particleboard, adhesives, and insulation materials.
- Methanol is increasingly used to produce olefins such as ethylene and propylene via the methanol-to-olefins (MTO) process (31%). These olefins are foundational in the production of plastics, synthetic rubber, and chemical fibers, with particularly rapid expansion in China.

Further details on chemical applications are provided in Sect. 7.2.

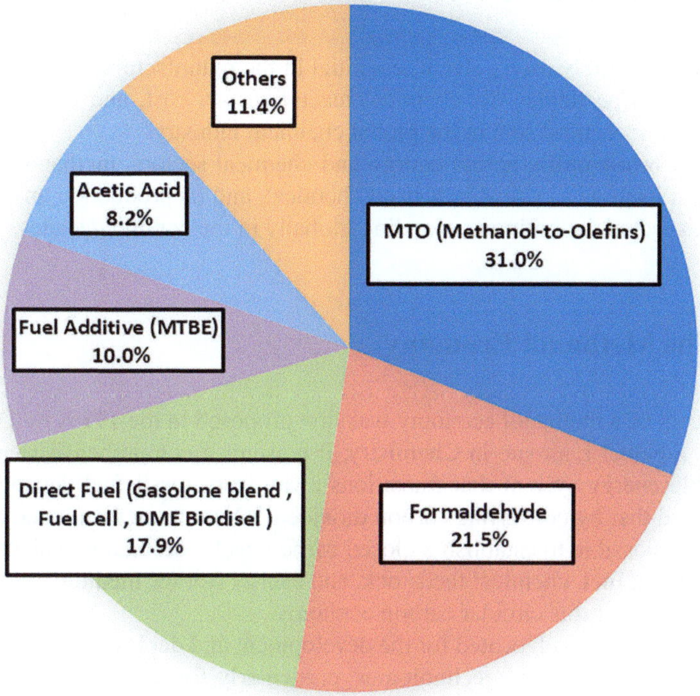

Fig. 5.3 Global methanol consumption in 2024 (data from Methanol Market Services Asia (MMSA)) [19]

Methanol as an Energy Source

One of the most important trends shown in Fig. 5.3 is that over one-third of global methanol production is now used as fuel. Its energy applications include:

- Fuel additive: Methanol is blended with gasoline to enhance octane number and combustion efficiency.
- Direct fuel: Used in internal combustion engines, especially in racing and commercial vehicles such as buses and trucks—with widespread deployment in China.
- Fuel cells: Utilized in Direct Methanol Fuel Cells (DMFCs), particularly for portable power sources and emergency backup systems. DMFCs offer high energy density and easy refueling, making them ideal for remote or mobile applications.
- Power generation: In non-electrified areas, methanol is also used as a fuel for electricity generation.

Other Industrial Uses

Methanol is a widely used solvent, essential in the manufacture of varnishes, inks, adhesives, and pesticides. Its chemical reactivity, low cost, and ease of handling contribute to its central role in the global chemical industry.

With its broad utility across energy and chemical sectors, methanol has firmly established itself as a critical industrial chemical, and its production and consumption are expected to continue expanding globally in the coming decades.

5.3 The Methanol Economy

The concept of a methanol economy was first proposed in the 1990s by Dr. George A. Olah, a Nobel Laureate in Chemistry. It presents a comprehensive vision for a sustainable energy system that transitions away from fossil fuel dependence. Dr. Olah argued that by converting carbon dioxide (CO_2) into methanol and reusing it, it would be possible to establish a closed carbon cycle. Methanol, which can function as a liquid fuel, chemical feedstock, and energy storage medium, was seen as a key component of this circular carbon economy.

Dr. Olah strongly advocated for the development and deployment of carbon capture and utilization (CCU) technologies, particularly for methanol production. He emphasized that methanol is feasible for global energy use because it can be stored and transported using existing infrastructure, such as tanks, pipelines, and ships.

In his seminal book [9], Dr. Olah discussed in detail the chemistry of methanol synthesis, including the production of dimethyl ether (DME), the conversion of methanol into olefins such as ethylene and propylene, the synthesis of gasoline, and the manufacture of formaldehyde. He also addressed the use of renewable energy and the production of hydrogen, while expressing skepticism about using hydrogen directly due to its storage and transport difficulties. Although not particularly opposed to nuclear power, he focused primarily on renewable sources such as solar, wind, geothermal, and ocean energy.

While Dr. Olah's methanol economy concept provides a broad foundation, it does not clearly define the strategic direction for international implementation. In contrast, this book seeks to present a quantitative and structured approach as described in the Introduction:

- Part I outlines the problem landscape.
- Part II discusses the technical and economic superiority of methanol as an energy carrier.
- Part III proposes action plans for industries, policymakers, businesses, and civil society to collaboratively build a methanol-centered energy system.

To reflect this strategic focus, we use the term "**Methanol Society**" in this book.

Real-World Implementations of the Methanol Economy

In 2012, *Carbon Recycling International (CRI) in Iceland* constructed the George Olah Renewable Methanol Plant [17], named in honor of Dr. Olah. This facility uses geothermal energy to convert captured CO_2 into methanol. As of May 2024, CRI produces approximately 214,000 tons of methanol annually from 310,000 tons of CO_2, a pioneering example of renewable methanol production. While the scale is not large relative to global demand, it is a milestone in sustainable chemical manufacturing.

In *Japan*, *Mitsui Chemicals* built one of the earliest plants utilizing factory-emitted CO_2 to synthesize methanol. Established in 2009, this facility produces 100 tons per year of methanol, representing an early application of CCU technology [18].

Another prominent approach is the production of *e-methanol*, which refers to methanol synthesized from green hydrogen (produced via electrolysis using renewable electricity) and captured CO_2. In *Sweden*, *Liquid Wind* is developing such systems, and in *Germany*, *BASF* has begun producing e-methanol at its *Ludwigshafen* facility. In 2024, BASF announced a strategic partnership with Envision Energy to scale up e-methanol production using primarily wind power.

Toward a Methanol Society

The ideal methanol society envisions methanol synthesized from green hydrogen and captured CO_2—including CO_2 emissions from fossil sources. However, this ideal cannot be achieved instantaneously. A pragmatic starting point is to use hydrogen derived from fossil fuels in combination with CO_2 capture, emphasizing that CO_2 capture is the immediate priority. A road map for transitioning to a methanol society will be presented in Chap. 9.

Currently, the largest methanol producer is **China** [17], where CO_2-to-methanol conversion technologies are rapidly advancing. One notable example is a facility in **Anyang City, Henan Province**, which began operations recently. This plant captures 160,000 tons of CO_2 per year and converts it into 110,000 tons of methanol, using emissions-to-liquids (ETL) technology developed by CRI. The CO_2 and H_2 used in this facility are derived from by-product gases generated during the coking process in steel production [20]. Emissions-to-liquids (ETL) technology involves catalytic conversion of CO_2 and H_2 into methanol, typically using Cu/ZnO-based catalysts at ~250 °C and 50–100 bar. The Anyang plant is one of the world's largest CCU-methanol facilities operating on industrial by-product gases, indicating strong industrial integration potential. This example illustrates how industrial by-products can be recycled into valuable chemicals, contributing to a more sustainable industrial ecosystem. Compared to traditional coal-based methanol synthesis, this process significantly reduces CO_2 emissions.

The operation of this facility supports China's carbon reduction goals and demonstrates the technical and economic feasibility of large-scale CO_2 conversion into useful chemicals. It also validates the broader concept of a methanol-based energy and chemical system, which could become a key pillar in future global sustainability strategies.

5.4 Stable Energy Supply

As discussed in Chap. 2, the largest single source of global CO_2 emissions is fossil fuel-based power generation (see Fig. 2.3). One promising approach to reducing these emissions is to replace fossil fuels with **methanol** or its derivative, **dimethyl ether (DME)**, as fuels for power plants. Both methanol and DME are well-suited for use in **thermal power generation**. In particular, when renewable energy sources are integrated into the power grid, these fuels can play a critical role as dispatchable or balancing power sources, helping to stabilize fluctuations in power supply and demand. The use of DME as a fuel will be discussed in more detail in Sect. 6.5. Here, we first review the fundamental principles of **power supply-demand balancing**.

As noted in Sect. 2.2, the integration of renewable energy introduces variability and intermittency into the power grid. Maintaining a continuous balance between electricity generation and consumption is essential to prevent grid instability, frequency deviations, or even blackouts. To manage this balance, electricity supply is typically categorized into three types:

i. Base load power
ii. Load-following (adjustment) power
iii. Auxiliary (backup) power

Base Load Power

Nuclear power plants are typically designed to provide base load power—a stable, continuous supply of electricity. These facilities operate most efficiently when running at a constant output, due to both technical constraints and economic considerations. Rapid output fluctuations can pose safety risks, and given that nuclear plants have high fixed costs and low variable costs, continuous full-capacity operation is economically preferable. However, newer nuclear technologies are emerging that enable load-following operation. These advanced reactors use improved control systems and fuel technologies to adjust output more flexibly in response to changes in demand. While this offers potential benefits, especially in grids with high shares of renewables, the adjustment capabilities of conventional nuclear plants remain limited.

Load-Following Power (Adjustment Power)

Adjustment power sources are critical to compensating for short-term fluctuations in power demand or supply. These facilities must be able to start up quickly and rapidly modulate output. Common examples include gas turbines and pumped storage hydropower. Such systems can respond within minutes, making them essential for real-time grid balancing, particularly during peak demand periods or renewable energy output fluctuations. By improving the flexibility and responsiveness of the grid, adjustment power sources support energy system resilience and contribute to long-term energy security.

Auxiliary Power (Backup Power)

Auxiliary power sources provide backup electricity during planned maintenance or unexpected outages of main power facilities. These systems include emergency diesel generators, battery energy storage systems (BESS), and other forms of standby generation. They are crucial for maintaining the operation of critical infrastructure such as hospitals, telecommunications, and emergency response facilities. Auxiliary power sources are also used to complement renewable generation, enhancing overall system stability and reliability. By reducing the risk of power disruptions and enabling smoother integration of renewables, they play an important role in ensuring a sustainable and robust energy supply.

Methanol and DME for Power Adjustment

Methanol combustion is relatively easy to start and stop, making it suitable for flexible power generation and grid stabilization. Its ability to respond to changes in demand makes it an attractive option for both load-following and peak-shaving applications. Methanol can also function as an auxiliary energy source, supporting system resilience during disruptions.

DME, which can also be combusted cleanly and efficiently, offers similar advantages. It allows for rapid startup and shutdown and thus can meet demand for fast-ramping power. Due to its higher combustion efficiency compared to methanol, DME may offer slight performance advantages in emergency peak-cutting and backup power scenarios.

5.5 Promotion of Commercialization by Existing Technology

One of the key advantages of methanol is that its use can be promoted immediately through the application of existing, mature technologies. To begin with, the hydrogenation of CO_2 to produce methanol is already a commercialized process. In current industrial chemical production, methanol is primarily synthesized from carbon monoxide (CO) and hydrogen (H_2). However, a small fraction of CO_2 is often included in the feedstock. The same catalyst system—typically Cu/ZnO/Al_2O_3—can be utilized for methanol synthesis from CO_2 as well. The Cu/ZnO-based catalyst used for methanol synthesis is relatively inexpensive, has a long operational lifespan, and can be regenerated, making it suitable even under current cost constraints for renewable hydrogen.

Comparison with Ammonia Synthesis

When comparing methanol synthesis to ammonia synthesis, the benefits of methanol become even more apparent. Ammonia production requires high temperatures (400–500 °C) and very high pressures (150–350 atm). These harsh conditions result in significant electricity consumption, accounting for approximately 1–2% of global power demand. An expansion of ammonia production would therefore greatly increase energy consumption, undermining its value as a sustainable energy carrier.

In contrast, methanol synthesis occurs under milder conditions—typically 250 °C and 50–100 atm. As a result, the energy demand of methanol production is considerably lower than that of ammonia.

Advantages in Hydrogen Regeneration

Moreover, when considering the reverse reaction—that is, regenerating hydrogen from methanol or ammonia—methanol again demonstrates a clear advantage. Methanol decomposition typically occurs at 200–300 °C, whereas ammonia cracking requires 400–500 °C. This temperature difference makes methanol more suitable for energy applications, as waste heat from hydrogen utilization systems can be effectively used to drive the reverse reaction for methanol.

Compatibility with Existing Infrastructure

Perhaps the most significant advantage of methanol is its immediate applicability as a fuel for internal combustion engines (see Fig. 7.1). As described in Chap. 6, methanol is already in use as a fuel for various types of engines, including methanol-powered cars and racing vehicles. Methanol is sometimes blended with biofuels or petroleum-based fuels to enhance performance.

DME, a derivative of methanol, is being used as a diesel substitute and cooking gas. DME emits lower levels of sulfur and nitrogen oxides, making it an appealing clean fuel. Another methanol derivative, methyl tert-butyl ether (MTBE), has been used as a gasoline additive to improve octane rating and combustion efficiency.

Crucially, these fuels can be integrated into existing fuel supply infrastructure. This stands in stark contrast to ammonia or hydrogen, which would require the development of entirely new supply chains and refueling stations. In contrast, methanol fueling stations can be deployed more cost-effectively using gasoline infrastructure.

Integration into the Chemical Industry

As elaborated in Chap. 7, methanol also enables the continuation and decarbonization of the current chemical industry. One of the most important developments is the production of ethylene and propylene from methanol via the methanol-to-olefins (MTO) process. These olefins are essential feedstocks for the manufacture of plastics. Importantly, waste plastic can be converted back into methanol, enabling a closed-loop carbon cycle in the chemical sector. This creates a realistic pathway toward a non-petroleum-based chemical industry. MTO technology is already being implemented at scale in China, demonstrating its technical feasibility using existing systems. By integrating plastic gasification, CO_2 capture, methanol synthesis, and the methanol-to-olefins (MTO) process, a closed-loop carbon recycling system can be established within the chemical industry.

This dual capability—supporting both energy and chemical production—is a distinctive feature of methanol that is not shared by other energy carriers. As discussed above, methanol offers a combination of advantages that make it uniquely suited for rapid commercialization:

- Established production technology
- Immediate applicability in energy systems
- Compatibility with existing infrastructure
- Integration with the existing chemical industry

No other compound offers this combination of technological maturity, versatility, and practicality. For this reason, methanol stands out as an exceptional candidate for accelerating the transition to a low-carbon energy and industrial system.

References

1. Kanai Y, Watanabe T, Fujitani T, Saito M, Nakamura J, Uchijima T (1994) Evidence for the migration of ZnOx in a Cu/ZnO methanol synthesis catalyst. Catal Lett 27:67–78
2. Fujitani T, Nakamura I, Watanabe T, Uchijima T, Nakamura J (1995) Methanol synthesis by the hydrogenation of CO_2 over Zn-deposited Cu(111) and Cu(110) surfaces. Catal Lett 35:297–302
3. Fujitani T, Nakamura I, Uchijima T, Nakamura J (1997) The kinetics and mechanism of methanol synthesis by hydrogenation of CO_2 over a Zn-deposited Cu(111) surface. Surf Sci 383:285–298
4. Fujitani T, Nakamura J (2000) The chemical modification seen in the Cu/ZnO methanol synthesis catalysts. Appl Catal 191:111–129
5. Nakamura J, Choi Y, Fujitani T (2003) On the issue of the active site and the role of ZnO in Cu/ZnO methanol synthesis catalysts. Top Catal 22:277–285
6. Nakamura J, Fujitani T, Kuld S, Helveg S, Chorkendorff I, Sehested J (2017) Comment on "active sites for CO_2 hydrogenation to methanol on Cu/ZnO catalysts". Science 357:6354
7. Quan J, Muttaqien F, Kondo T, Kozarashi T, Mogi T, Imabayashi T, Hamamoto Y, Inagaki K, Hamada I, Morikawa Y, Nakamura J (2019) Vibration-driven reaction of CO_2 on Cu surfaces via Eley–Rideal-type mechanism. Nat Chem 11(8):722–729
8. Takeyasu K, Sawaki Y, Imabayashi T, Putra SEM, Halim HH, Quan J, Hamamoto Y, Hamada I, Morikawa Y, Kondo T, Fujitani T, Nakamura J (2022) Hydrogenation of Formate species using atomic hydrogen on a Cu(111) model catalyst. J Am Chem Soc 144:12158–12166
9. Olah GA, Goeppert A, Surya Prakash GK (2006) Beyond oil and gas: the methanol economy. Wiley-VCH
10. Olah GA (2005) Beyond oil and gas : the methanol economy. Angew Chem Int Ed 44:2636–2639
11. Olah GA, Goeppert A, Surya Prakash GK (2009) Chemical recycling of carbon dioxide to methanol and dimethyl ether: from greenhouse gas to renewable, environmentally carbon neutral fuels and synthetic hydrocarbons. J Org Chem 74:487–498
12. Olah GA (2013) Towards oil Independence through renewable methanol chemistry. Angew Chem Int Ed 52:104–107
13. Goeppert A, Czaun M, Jones JP, Surya Prakash GK, Olah GA (2014) Recycling of carbon dioxide to methanol and derived products–closing the loop. Chem Soc Rev 43:7995–8048
14. Bockris JO'M (2010) Would methanol formed from CO_2 from the atmosphere give the advantage of hydrogen at lesser cost? Int J Hydrog Energy 35:5165–5172
15. Bockris JO'M (2011) Is methanol using CO_2 from the atmosphere a new fuel to replace gasoline? Smart Grid Renew Energy 2:152–157
16. Bockris JO'M (2013) The hydrogen economy: its history. Int J Hydrog Energy 38:2579–2588
17. Home page of Carbon Recycling International. https://carbonrecycling.com/
18. Mitsui Chemicals Inc Report. https://jp.mitsuichemicals.com/en/release/2008/080825e.htm
19. Homepage of Methanol Institute (2024) Methanol Price, Data from MMSA. https://www.methanol.org/methanol-price-supply-demand/
20. World's largest CO_2-to-methanol plant starts production. https://www.hydrocarbonprocessing.com/news/2022/10/worlds-largest-co2-to-methanol-plant-starts-production

Chapter 6
Energy Use of Methanol

Abstract Methanol offers a promising solution to the challenges of hydrogen storage, transport, and utilization. This chapter focuses on methanol as a hydrogen carrier, emphasizing its chemical composition (CH_3OH = 12.5% hydrogen by weight) and ease of reformation into hydrogen gas. Compared to compressed or liquefied hydrogen, methanol is safer, denser, and easier to handle. This chapter explains reforming methods such as steam reforming, partial oxidation, and autothermal reforming and quantifies their energy efficiency and CO_2 emissions. It also discusses integration with fuel cells, especially proton exchange membrane (PEMFC) and solid oxide (SOFC) types, highlighting methanol's advantages for distributed energy systems and backup power. This chapter explores practical use cases such as portable power devices, off-grid systems, and heavy transport. Methanol's potential to leverage existing fuel infrastructure makes it a compelling candidate for a hydrogen-based energy future.

Keywords Hydrogen carrier · Methanol reforming · Fuel cells · PEMFC · SOFC · Hydrogen storage · Distributed energy

6.1 Superiority as an Energy Carrier

As noted in Part I, global efforts are underway to develop a hydrogen economy, wherein hydrogen is produced using electricity from solar, wind, and geothermal sources. However, direct storage and transport of hydrogen face significant challenges due to its physical properties. Liquefaction, for instance, requires cryogenic temperatures and results in energy losses of up to 40–45 MJ/kg.

For these reasons, attention is shifting toward hydrogen carriers—molecules in which hydrogen is chemically stored for easier handling. Among these, **methanol** is the most promising candidate.

J. Nakamura, M. Ishida, *Blueprint for a Methanol Society*,
https://doi.org/10.1007/978-981-95-3465-4_6

Thermochemical Reactions Involving Methanol

Methanol is synthesized from 1 mol of CO_2 and 3 mol of H_2 via the following exothermic reaction:

$$3H_2 + CO_2 \rightleftharpoons CH_3OH(l) + H_2O(l) \quad \Delta_r H^\circ = -131.0\,kJ/mol - CH_3OH \tag{6.1}$$

This enthalpy value applies when both methanol and water are in liquid form; for gas-phase products, the value is −49.4 kJ/mol. Although water might appear as a by-product, it can be reformed back into hydrogen via the methanol steam reforming reaction [1]:

$$CH_3OH(l) + H_2O(l) \rightleftharpoons 3H_2 + CO_2 \quad \Delta_r H^\circ = 131.0\,kJ/mol - CH_3OH \tag{6.2}$$

This reaction is endothermic, meaning the input energy ideally matches the output in reverse. The reforming process operates at 250–300 °C, and if waste heat is available at this temperature, it can drive the reaction efficiently. In this way, the system acts as a chemical heat pump [2, 3].

Combustion of methanol releases energy as follows:

$$CH_3OH(l) + \frac{3}{2}O_2 \rightleftharpoons 2H_2O(l) + CO_2 \quad \Delta_r H^\circ = -726.5\,kJ/mol - CH_3OH \tag{6.3}$$

The combustion of 3 mol of H_2, obtained from methanol reforming, yields:

$$3H_2 + \frac{3}{2}O_2 \rightleftharpoons 3H_2O(l) \quad \Delta_r H^\circ = -857.4\,kJ/mol - 3H_2 \tag{6.4}$$

By Hess's Law, this energy difference accounts for the earlier heat input in the endothermic step. This demonstrates that methanol reforming can effectively reproduce the flame temperature and exergy of hydrogen combustion using moderate-temperature waste heat.

In theoretical terms, direct conversion of hydrogen to methanol retains 83.0% of the energy: (726.5/857.4) × 100 = 84.7%. However, through reforming, nearly the full energy content of hydrogen can be recovered.

Comparison with Methane

Another hydrogen carrier is methane, synthesized as:

$$4H_2 + CO_2 \rightleftharpoons CH_4 + 2H_2O(l) \quad \Delta_r H^\circ = -252.9\,kJ/mol - CH_4 \tag{6.5}$$

Compared to hydrogen (1143.3 kJ for 4 mol), methane retains about **78%** of the energy.

$$CH_4 + 2O_2 \rightleftharpoons CO_2 + 2H_2O(l) \quad \Delta_r H^\circ = -890.4\,\mathrm{kJ/mol} - CH_4 \tag{6.6}$$

However, steam reforming of methane requires much higher temperatures (700–800 °C) than methanol, limiting the efficient use of waste heat. Although LNG infrastructure exists for methane, methanol is thermally superior for distributed reforming.

CO_2 Procurement Pathways for Methanol Synthesis

Since methanol production requires CO_2, the method of **CO_2 sourcing** is crucial. The following four models shown in Fig. 6.1 can be envisioned:

i. **CO_2 Emission Capture System**
   CO_2 emitted during the use of fossil fuels is captured and mixed with green hydrogen to synthesize methanol. CO_2 is ultimately emitted into the atmosphere

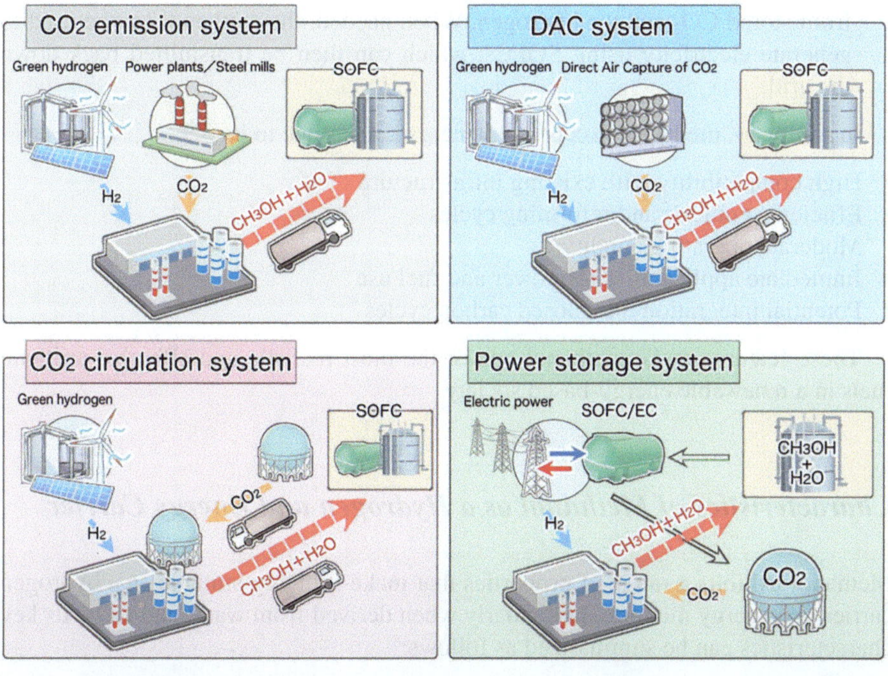

Fig. 6.1 CO_2 procurement pathways for methanol synthesis. (i) CO_2 emission capture system, (ii) direct air capture (DAC) system, (iii) CO_2 circulation system, and (iv) power storage system

by the SOFC or internal combustion engine that uses methanol. Although CO_2 is not substantially reduced, renewable energy is used, so no CO_2 is emitted during the energy use process. However, the cost of CO_2 capture and hydrogen production remains high.

ii. **Direct Air Capture (DAC) System**
   CO_2 is sourced from ambient air via DAC technology, combined with green hydrogen. DAC is actively being researched and could enable net-negative emissions, though current cost and energy demands are significant. Pairing DAC with renewable electricity is a possible solution.

iii. **CO_2 Circulation System**
   The methanol synthesis plant is the center, and after transporting the methanol to the power generation site (SOFC or internal combustion engine) and using the energy, the captured CO_2 is transported back to the methanol synthesis plant. Hydrogen derived from renewable energy is transported to the methanol synthesis plant. A distribution network using liquefied or solid CO_2 (e.g., dry ice) is already established. This method is seen as technically feasible and economically attractive, though further refinement is needed for large-scale energy applications.

iv. **Power Storage System**
   Methanol and CO_2 can be used as chemical energy storage media, forming a closed-loop system. Using the grid, hydrogen is produced by electrolysis of water using electricity generated elsewhere, and methanol is then synthesized from stored CO_2 and the hydrogen. When needed, the methanol can be used to generate electricity using SOFCs, which can then be transmitted back down the grid.

In summary, methanol excels as an energy carrier due to its:

- High compatibility with existing infrastructure
- Efficient synthesis and reforming cycles
- Moderate operating conditions
- Immediate applicability for power and fuel use
- Potential integration into closed carbon cycles

These features make methanol one of the most realistic and effective artificial fuels in a renewable energy-based society.

Characteristics of Methanol as a Hydrogen and Energy Carrier

Methanol exhibits a range of properties that make it highly suitable as a hydrogen carrier and energy medium, particularly when derived from water and CO_2. Its key characteristics can be summarized as follows:

- Low energy density (726.5 kJ/mol) and the presence of water in the mixture reduce the risk of explosion. In Japan, methanol-water solutions with concentrations below 60% are not classified as hazardous materials.

- It is widely used in commercial products, such as fuel alcohol and windshield washer fluids, indicating social acceptance and infrastructure compatibility.
- The methanol steam reforming temperature is relatively low (200–300 °C), enabling effective utilization of waste heat.
- Hydrogen derived from water can be regenerated via reforming: 857.4 kJ/ H_2 −3mol.
- In high-temperature fuel cells (e.g., solid oxide fuel cells, SOFCs), internal reforming of methanol can be integrated, improving overall power generation efficiency.
- High pressures can be achieved easily and efficiently using liquid pumps and external vaporization with thermal expansion.
- Methanol is resistant to carbon deposition, and excess water (beyond the stoichiometric steam/carbon (S/C) ratio) can be recycled in the system.
- Methanol synthesis is feasible at any scale, provided H_2 and CO_2 are available, allowing for decentralized production.
- Methanol exhibits excellent stability under long-term storage in sealed containers at ambient temperature, unlike fuels such as gasoline or kerosene, which degrade over time.
- It can be stored, transported, and converted using standard materials and infrastructure, unlike liquid hydrogen (LH_2) or ammonia (NH_3) which requires specialized systems.
- The supporting technologies for CO_2 separation, capture, liquefaction (or solidification), storage, and transport are already well established and industrially implemented.
- Methanol's ability to be synthesized on-site in small modular systems allows for decentralized energy solutions in remote or off-grid locations.

Limitations

The disadvantages of methanol are relatively limited, but they include the followings:

- It is highly volatile and evaporates quickly when exposed to air.
- Toxic if ingested, methanol can cause severe health effects, including blindness and death, even at relatively small doses.

Strategic Advantages

The fact that methanol is a liquid at ambient temperature is a major practical advantage. Compared to other hydrogen carriers such as ammonia or liquid hydrogen, methanol offers a superior combination of low handling costs, moderate synthesis and reforming conditions, and compatibility with existing infrastructure.

If methanol can be directly supplied to a stationary power generation unit, such as an SOFC, it enables a simple, compact, and highly efficient system configuration. This makes methanol especially attractive for special-purpose power sources, including:

- Off-grid and remote locations.
- Unmanned aerial vehicles (UAVs).
- Backup and emergency power systems.

In conclusion, methanol combines practical handling, energy versatility, and technological maturity, making it an ideal candidate for widespread adoption as a sustainable energy carrier.

6.2 Methanol for Internal Combustion Engines

Methanol is already being utilized in internal combustion engines, and its use is expected to expand due to its environmental benefits, combustion characteristics, and compatibility with existing infrastructure. Methanol can be used:

- As a **blending component** with gasoline (used in significant proportions to form part of the fuel itself).
- As a **gasoline additive** (added in small amounts to improve fuel properties such as water tolerance or cold start performance).
- As a **standalone fuel** in dedicated engines.

Because of its favorable fuel properties, full-scale commercialization of methanol-powered engines is anticipated.

Advantages and Characteristics

Methanol has a very high-octane number, higher than that of gasoline, making it ideal for spark ignition engines. Methanol's flame speed and heat of vaporization are higher than gasoline, which contributes to better knock resistance and cooling effects during combustion. Its low cetane number is less suited to compression-ignition systems but contributes to knock resistance and efficient combustion. Compared to diesel and gasoline, methanol combustion:

- Produces **fewer harmful emissions** (e.g., NO_x and particulate matter).
- Contains **no sulfur or nitrogen atoms**, reducing pollution.
- Contains **oxygen**, promoting more complete combustion.

These factors make methanol a clean alternative fuel that can help mitigate urban air pollution.

However, methanol's lower energy density (about half that of diesel by weight) means that vehicles require larger fuel tanks or more frequent refueling to maintain the same driving range.

Applications Beyond Road Vehicles

Methanol is also being applied in other sectors:

- Thermal power generation, particularly small-scale distributed systems, where its clean-burning nature and operability are advantages.
- Backup generators and remote area power supply, where liquid fuel is preferred for ease of handling.
- As a reducing agent in the steelmaking process, where methanol can replace coke, reducing CO_2 emissions.
- In the cement and petrochemical industries, where it may substitute for fossil fuels, and where oxygen-enriched combustion can facilitate CO_2 capture.
- Near-engine methanol reformers can generate hydrogen on-site for either combustion or fuel cell use (e.g., SOFC). Whether to reform methanol or directly capture CO_2 depends on cost and system efficiency.

Automotive Sector and Global Trends

Methanol's high-octane value also helps prevent engine knocking, promoting cleaner combustion. The cost of methanol refueling stations is comparable to gasoline/diesel infrastructure and far cheaper than hydrogen stations (which cost over US$ two million each) or LNG stations gaining attention in Europe.
China is currently at the forefront of methanol vehicle development as part of its strategy to reduce oil imports and improve environmental performance. Chinese automaker Geely has developed several methanol-powered vehicle models [4], priced competitively with gasoline vehicles. Geely has also invested in Carbon Recycling International (CRI) in Iceland, a company specializing in CO_2-to-methanol conversion. The Chinese government is actively supporting pilot programs, offering consumer subsidies and funding station conversions to promote methanol fuel usage. While China leads in methanol vehicle deployment, Europe and the United States are exploring its use primarily in shipping and fuel cells due to stricter emission regulations and alternative technology preferences. This is part of a broader national strategy for fuel diversification and energy security.

Outside China, companies like Blue World Technologies (Denmark) and Karma Automotive (United States) are exploring methanol fuel cell systems. Karma is developing methanol-based fuel cell electric vehicles (FCEVs), offering short refueling times and simple liquid fuel logistics similar to gasoline.

CO_2 Neutrality Considerations

Capturing CO_2 emissions from individual methanol vehicles is not currently cost-effective. However, releasing CO_2 into the atmosphere does not necessarily mean that it does not contribute to carbon neutrality. Synthesizing methanol from H_2 produced using renewable energy and captured CO_2 emissions reduces net CO_2 emissions because of the renewable energy. The significance of using methanol in vehicles is that it helps raise public awareness of the use of methanol. Because automobiles are closely related to civic activities, it will help citizens understand the possibilities of a methanol society.

Methanol in Maritime Transport

Methanol is gaining attention as a marine fuel due to its:

- Low sulfur content, enabling compliance with IMO emissions regulations.
- Lower infrastructure costs than LNG.
- Ease of storage and handling.

Companies like **Wärtsilä (Finland)** offer a range of **methanol engines** and **conversion kits** for existing ships (see Fig. 6.2). **Rolls-Royce (UK)** and **MAN Energy Solutions (Germany)** are also producing methanol marine engines.

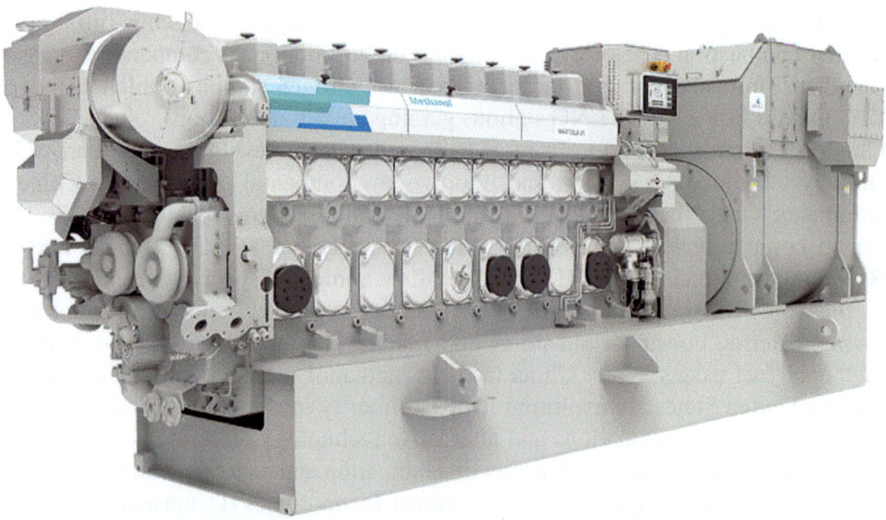

Fig. 6.2 Wärtsilä methanol engines [5], Photo courtesy of Wärtsilä Corporation

Technical Challenges

Despite its advantages, methanol use in internal combustion engines faces several technical challenges:

i. Corrosiveness
Methanol is chemically aggressive toward materials like aluminum, copper, and certain rubbers. Special materials and coatings are needed for fuel lines, injectors, and seals.
ii. Cold Startability
Methanol's lower volatility compared to gasoline can cause starting problems in cold weather, due to poor vaporization and mixture formation.
iii. Low Energy Density
With lower energy per unit volume than gasoline or diesel, larger tanks or more frequent refueling are required.
iv. Toxicity
Methanol is toxic if ingested and irritating to skin and eyes, requiring careful handling and appropriate safety protocols.
v. Infrastructure Development
Methanol requires dedicated refueling infrastructure. While this is a cost, it remains significantly lower than that of hydrogen or LNG systems.

Outlook

Despite these challenges, methanol-fueled internal combustion engines are already viable with existing technology. While methanol fuel cells are still under development and commercialization, expanding the use of methanol in engines represents a realistic and near-term path for scaling up methanol energy utilization.

6.3 Methanol for Solid Oxide Fuel Cells (SOFC)

Fuel cells are electrochemical devices that convert chemical energy directly into electrical energy. While hydrogen is the most commonly used fuel, various other fuels—including methanol—can also be utilized depending on the type of fuel cell. The basic components of a fuel cell include an anode, a cathode, and an electrolyte. Since fuel is continuously supplied, fuel cells are often considered more akin to generators than batteries.

The main types of fuel cells, categorized by their operating temperatures and applications, are as follows:

- **Proton Exchange Membrane Fuel Cells (PEMFC):** Operate at approximately 80 °C; they offer rapid startup and are widely used in transportation and residential systems.
- **Phosphoric Acid Fuel Cells (PAFC)**: Operate at around 200 °C; they are known for high reliability and are suitable for commercial-scale stationary power generation.
- **Molten Carbonate Fuel Cells (MCFC):** Operate at about 650 °C; capable of utilizing CO_2 as part of the electrochemical reaction, these are typically used for large-scale power plants.
- **Solid Oxide Fuel Cells (SOFC):** Operate at high temperatures ranging from 700 °C to 1000 °C; they can accept a wide variety of fuels, including natural gas, biogas, and methanol, and are well-suited for industrial and distributed power applications.

Fuel cells are characterized by high efficiency. Even with electrical generation efficiencies in the range of 50–70%, the overall system efficiency can be significantly increased through the utilization of waste heat, such as in combined heat and power (CHP) systems. This thermal energy results primarily from electrochemical losses, including overpotentials due to activation energy barriers. Effectively recovering and utilizing this heat is essential to realizing the full energy potential of the fuel cell system.

In addition to their high efficiency, fuel cells emit very low levels of pollutants and operate quietly, making them environmentally and socially advantageous. Their scalability—from small residential units to large industrial systems—adds to their versatility.

Methanol, in particular, is a promising fuel for SOFCs [6–9]. It offers a high volumetric energy density, is easier to store and transport than hydrogen, and can be reformed internally within the high-temperature environment of the SOFC. This internal reforming capability reduces the need for external fuel processing systems, potentially simplifying system design and lowering overall costs [6, 7]. As a liquid at ambient conditions, methanol also benefits from established infrastructure for handling and distribution, making it a practical option for stationary power generation and future low-carbon energy systems.

General Challenges of Fuel Cells and Future Outlook:

- High initial costs due to expensive membrane materials and catalysts in PEMFC such as platinum.
- Limited hydrogen infrastructure, including production, storage, and distribution systems.
- Durability and long-term performance concerns under real-world operating conditions.

Despite these challenges, fuel cells are expected to play a crucial role in future energy systems, particularly in a methanol-based energy society. Readers are referred to further literature for more detail [10–12].

Among the various types of fuel cells, solid oxide fuel cells (SOFCs) are particularly well-suited for using methanol as a fuel, as shown in Fig. 6.3. Their high

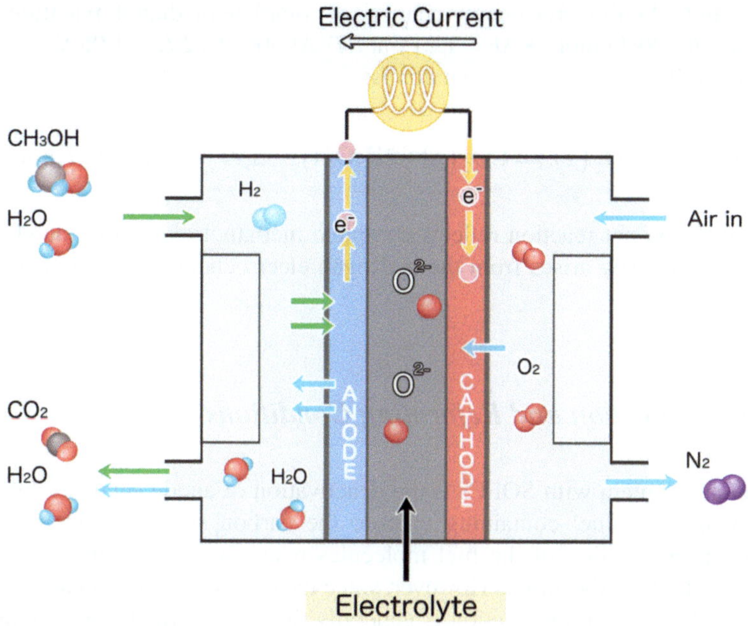

Fig. 6.3 Schematic structure of a solid oxide fuel cell (SOFC)

operating temperature enables internal reforming, allowing methanol to be converted into hydrogen directly within the cell.

Internal Reforming in SOFCs

An SOFC typically consists of three layers:

- **Anode**: where fuel (e.g., hydrogen) is oxidized.
- **Cathode**: where oxygen is reduced.
- **Electrolyte**: a solid oxygen-ion conductor, typically yttria-stabilized zirconia (YSZ).

When methanol is used as a fuel, hydrogen is not supplied externally but is generated in situ via the endothermic steam reforming reaction of Eq. 6.2. This hydrogen then participates in the electrochemical redox reactions:

$$\text{Anode: } 3H_2 + 3O^{2-} \rightarrow 3H_2O + 6e^- \quad (6.7)$$

$$\text{Cathode: } \frac{3}{2}O_2 + 6e^- \rightarrow 3O^{2-} \quad (6.8)$$

The standard Gibbs free energy change for complete methanol oxidation is:
$\Delta G = -702.8\,\text{kJ/mol} \rightarrow \Delta E = 1.21\,\text{V at } RT$ At 600 °C, $\Delta E \approx 1.05$ V.
The overall electrochemical reaction is:

$$CH_3OH(l) + \frac{3}{2}O_2(g) \rightleftharpoons CO_2(g) + 2H_2O(l), \quad \Delta_r H^\circ = -726.5\,\text{kJ/mol} \tag{6.9}$$

Although the net reaction reflects complete methanol oxidation, the electricity generation primarily arises from the hydrogen electrochemical reactions (Eqs. 6.7 and 6.8).

Carbon Deposition and Reforming Conditions

A technical problem with SOFCs is the deactivation of anode catalysts due to carbon. When using fuel containing carbon, the carbon species generated by the decomposition reaction of the fuel molecules reacts with oxide ions that diffuse from the cathode to the anode. However, since the deposition rate of carbon species due to the decomposition reaction is generally faster than the diffusion rate of the oxide, the deposited carbon species accumulate and cover the catalyst surface where the reaction is taking place. This is catalyst poisoning due to carbon deposition. When methanol-water is used as fuel, H_2O reacts with the carbon species on the surface and is oxidized to CO and CO_2, significantly suppressing carbon deposition. Internal reforming can also help suppress carbon deposition, particularly when the steam-to-carbon ratio is maintained above critical thresholds.

- Carbonaceous species may form from partial fuel decomposition at the anode.
- The presence of steam (H_2O) promotes their oxidation to CO or CO_2, thus cleaning the anode surface.

To prevent degradation, a steam-to-carbon ratio (S/C) ≥ 1.5 is generally recommended. Experimental studies have shown that power output remains stable at operating temperatures of 750 °C or higher, and the S/C ratio has minimal impact on power output as long as methanol conversion is complete.

Experimental Results of Methanol-Fueled SOFC Power Generation

Figure 6.4 presents the experimental results of SOFC performance using methanol at different steam-to-carbon (S/C) ratios—1.0, 1.5, and 2.0—and reaction temperatures of 700 °C, 750 °C, and 800 °C. Here, "carbon" refers to the carbon atoms contained in methanol (CH_3OH).

In these experiments:

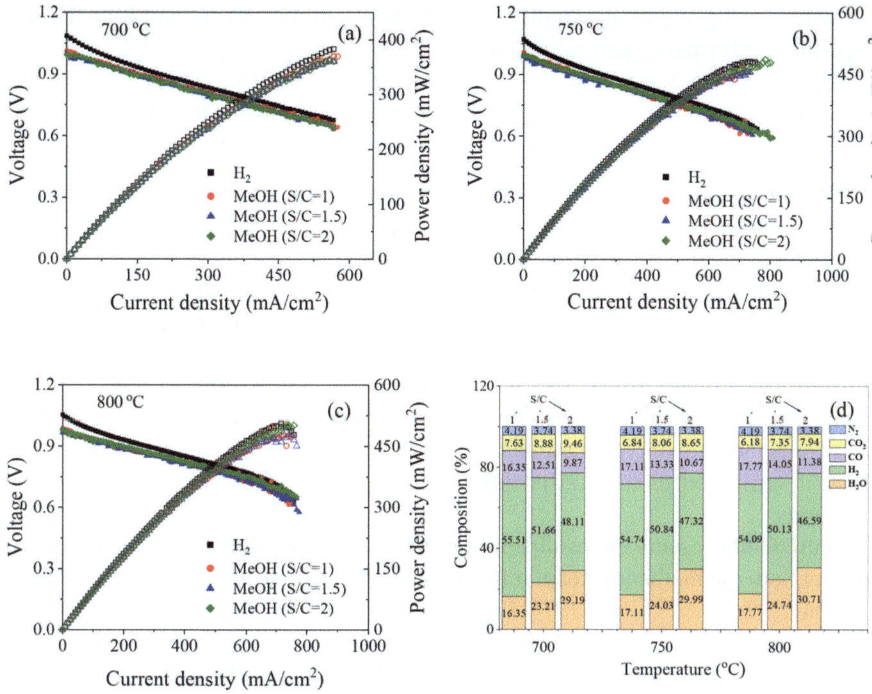

Fig. 6.4 Performance of methanol-fueled SOFC under varying operating conditions: (**a**–**c**) Current–voltage (I–V, closed symbols) and current–power density (I–P, open symbols) curves at different steam-to-carbon (S/C) ratios under operating temperatures of (a) 700 °C, (**b**) 750 °C, and (**c**) 800 °C. (**d**) Equilibrium gas-phase compositions calculated thermodynamically as a function of S/C ratio and temperature. Reprinted from [9] with the permission of American Chemical Society

- A Ni/YSZ (nickel/yttria-stabilized zirconia) catalyst was employed in the anode region, serving as a pre-reformer.
- Hydrogen was presumed to be generated via the methanol steam reforming reaction (Eq. 6.2).
- Without the reforming catalyst, significant carbon deposition and performance degradation occurred at S/C = 1.0.
- When the reforming catalyst was present, methanol conversion efficiency at S/C = 1.0 improved to approximately 95%, and degradation was mitigated, highlighting the critical role of pre-reforming in carbon management.

Effect of Temperature and S/C Ratio on Power Output

- At 700 °C, the power density reached ~400 mW/cm^2 at a current density of 600 mA/cm^2.

- Power density increased slightly to ~450 mW/cm^2 at 750 °C and 800 °C, suggesting performance saturation above 750 °C.
- The influence of the S/C ratio was relatively minor, as shown in Fig. 6.4a–c. This is attributed to high methanol utilization efficiency under all tested S/C conditions, which minimized the sensitivity of power output to the S/C ratio.

With the incorporation of the reforming catalyst, the methanol conversion efficiency at an S/C ratio of 1.0 improved to approximately 95%, demonstrating the effectiveness of catalyst-assisted internal reforming.

Regarding the effect of reaction temperature, the power density reached approximately 400 mW/cm^2 at 600 mA/cm^2 and 700 °C and increased to around 450 mW/cm^2 at both 750 °C and 800 °C. This plateau in performance suggests that further temperature increases above 750 °C offer limited benefit in terms of power output.

The influence of the S/C ratio on power generation is illustrated in Fig. 6.4a–c. The results indicate only a marginal effect of varying the S/C ratio on output power, which is attributed to the consistently high methanol utilization across all tested conditions.

Figure 6.4d presents the results of thermodynamic equilibrium calculations considering both the methanol steam reforming and the reverse water-gas shift (RWGS) reaction:

$$CO_2 + H_2 \rightleftharpoons CO + H_2O(1), \quad \Delta_r H^\circ = -2.8\,\text{kJ/mol} \tag{6.10}$$

These calculations support the experimental observation that the equilibrium concentrations of H_2 and CO—both serving as key fuels for SOFCs—are relatively insensitive to changes in the S/C ratio. As a result, the electrical power output remains largely unaffected by increasing S/C.

Based on these results, an especially noteworthy outcome is that an S/C ratio of 1.0 appears optimal. This is because 1 mol of CO_2 and 3 mol of H_2 react stoichiometrically to form 1 mol of methanol and 1 mol of water, making it possible to feed the SOFC directly with a methanol–water solution in a 1:1 molar ratio, without additional adjustment. Moreover, operation at S/C = 1.0 effectively mitigates degradation phenomena such as carbon deposition.

Collectively, these findings demonstrate both the technical feasibility and practical potential of SOFC systems operating with internally reformed methanol as a viable fuel.

System Concept and Heat Integration

A conceptual solid oxide fuel cell (SOFC) system utilizing methanol as fuel (Fig. 6.5) typically includes the following components:

- **Pre-reformer** (optional, for partial conversion of methanol prior to SOFC inlet).

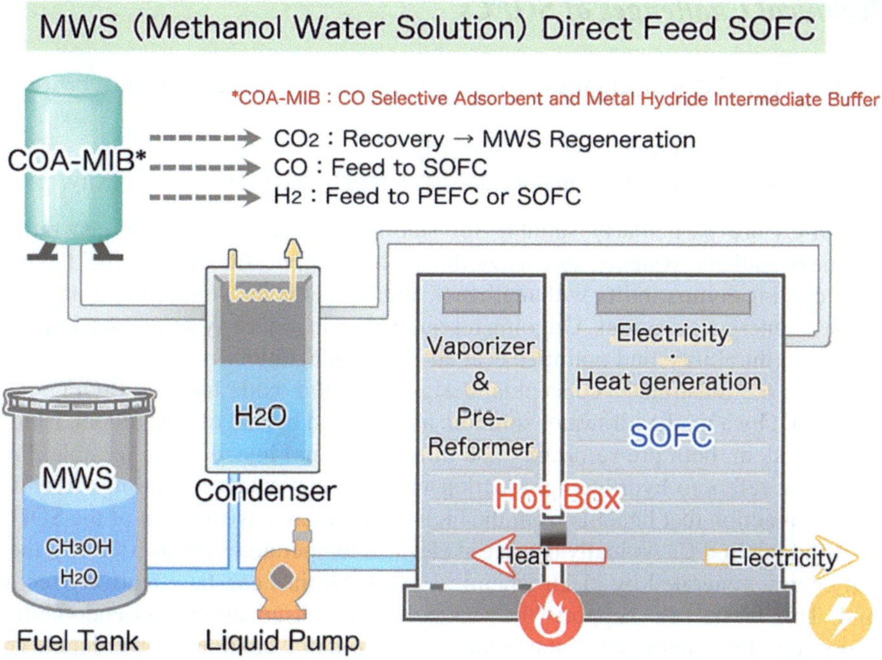

Fig. 6.5 Conceptual design of a solid oxide fuel cell (SOFC) system fueled by methanol–water solution (MWS). The system integrates pre-reforming, internal reforming, and thermal recovery to enable high-efficiency power generation. Exhaust gases (CO, H_2, and CO_2) are selectively recovered and reused to enhance system performance and support a closed carbon cycle. Proposed by Ishida

- **Internal reforming chamber** (within the SOFC anode).
- **SOFC stack.**
- **Heat recovery units** for vaporizing and preheating the methanol–water solution.
- **Exhaust gas treatment units** for recovering unreacted CO, H_2, and CO_2 for recycling.
- **Exhaust gas treatment**: recovery of CO, H_2, and CO_2 for reuse.

System Advantages

- Simplified and compact system configuration.
- High overall energy efficiency through internal heat utilization.
- Robust operation with fewer external subsystems.

Technical Challenges of SOFCs

- High operating temperatures (700–1000 °C) require thermally stable, corrosion-resistant materials.
- Long startup times due to thermal inertia.
- High initial capital costs, mainly due to materials and system complexity.

SOFCs are particularly suitable for stationary distributed power generation, marine propulsion systems, and heavy-duty vehicle applications. Their fuel flexibility, including compatibility with methanol, and high efficiency make them attractive for clean energy transitions. Ongoing research aims to reduce system costs, improve long-term durability, and optimize heat and mass integration.

Figure 6.5 illustrates a conceptual design of a solid oxide fuel cell (SOFC) system fueled by a methanol–water solution, as proposed by one of the authors, Ishida. In this system, both pre-reforming and direct methanol injection are feasible. Pre-reforming refers to hydrogen production via methanol steam reforming, an endothermic reaction that benefits from the high-temperature environment of the SOFC.

Although SOFCs typically operate at elevated temperatures (800–1000 °C), thermal energy generated by electrochemical overpotentials can be effectively recovered. This waste heat supports the vaporization, pressurization, and endothermic reforming of the methanol–water solution with minimal auxiliary power input. The resulting high-temperature, high-pressure hydrogen-rich gas is then fed into the SOFC. Such thermal integration is essential to maximizing the system's energy utilization efficiency, especially when using liquid energy carriers like methanol.

Exhaust gas utilization further enhances system efficiency. Major exhaust components include CO_2, H_2O, CO, and H_2. Among them, CO and H_2 can be selectively separated using CO-selective adsorbents and metal hydride intermediate buffers. The recovered CO may be recirculated into the SOFC, while H_2 can be redirected to either a polymer electrolyte fuel cell (PEFC) or a secondary SOFC unit. In addition, the system enables recovery of CO_2, which can be recycled as a carbon source for methanol synthesis—effectively contributing to a closed carbon loop.

A key advantage of this system lies in its relatively simple configuration, cost-effectiveness, and operational robustness, making it a strong candidate for practical deployment.

However, several technical challenges remain. The high operating temperatures of SOFCs require advanced materials with thermal and chemical stability, contributing to high initial capital costs. In addition, their thermal inertia leads to long startup times, making them less suitable for applications requiring rapid power availability.

Despite these limitations, SOFCs are considered highly promising for stationary power generation, distributed energy systems, marine propulsion, and heavy-duty vehicles. Thanks to their fuel flexibility and high efficiency, SOFCs are positioned as a key technology for the future of clean energy, and continued research aims to address remaining technical hurdles and support commercial viability.

6.4 Direct Methanol Fuel Cell (DMFC)

The **Direct Methanol Fuel Cell (DMFC)** is a type of fuel cell that directly utilizes methanol (CH_3OH) as a fuel and generates electricity through the electrochemical oxidation of methanol and water (H_2O). The basic structure of the DMFC consists of an anode, a cathode, and a proton exchange membrane (PEM) as shown in Fig. 6.6. At the anode, methanol and water are oxidized by a catalyst (typically a platinum-ruthenium alloy) to produce carbon dioxide (CO_2), protons (H^+), and electrons (e^-). The protons pass through the PEM, while the electrons flow through the external circuit to reach the cathode, where they combine with oxygen (O_2) and protons to produce water (H_2O). The PEM functions as a separator between the anode and cathode, selectively allowing only protons to pass through.

The main electrochemical reactions occurring in the DMFC are as follows:

- **Anode reaction:**

$$CH_3OH + H_2O \rightarrow CO_2 + 6H^+ + 6e^- \tag{6.11}$$

- **Cathode reaction:**

$$\frac{3}{2}O_2 + 6H^+ + 6e^- \rightarrow 3H_2O \tag{6.12}$$

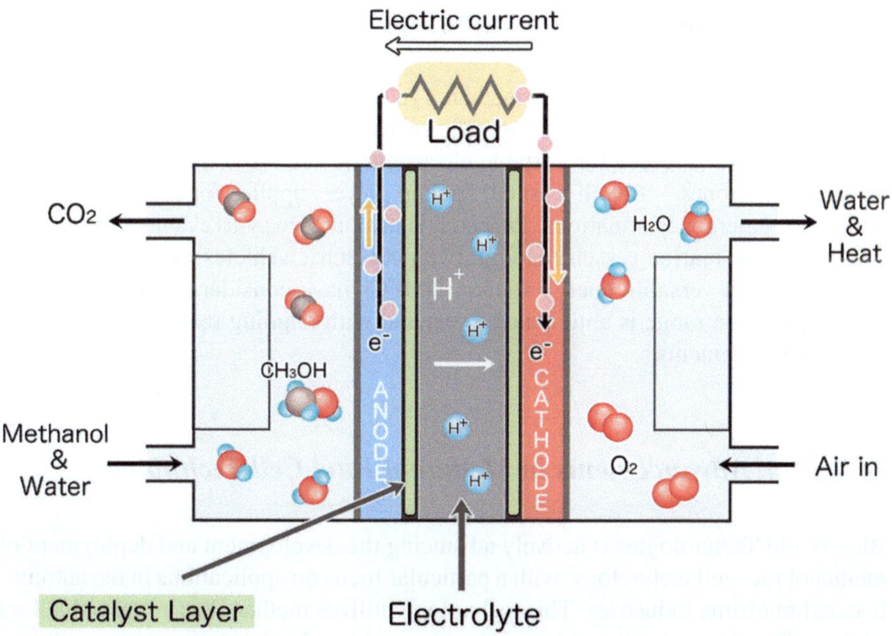

Fig. 6.6 Direct methanol fuel cell

- **Overall cell reaction:**

$$CH_3OH + \frac{3}{2}O_2 \rightarrow CO_2 + 2H_2O \tag{6.13}$$

The DMFC offers several advantages, including the simplicity of fuel supply, high energy density, and operation at relatively low temperatures. Methanol is a liquid fuel, making it easy to store and handle, and its relatively high energy density enables extended operational periods. The volumetric energy density of methanol (about 15.6 MJ/L) is significantly higher than that of compressed hydrogen (4–5 MJ/L at 700 bar), making it advantageous for compact applications. Furthermore, DMFC systems typically operate at moderate temperatures (60–120 °C), simplifying system design and reducing weight compared to high-temperature fuel cells.

However, the DMFC also presents several challenges. A significant issue is **methanol crossover**, wherein methanol permeates through the PEM to the cathode side, leading to a reduction in fuel cell efficiency and potential poisoning of the cathode catalyst. Additionally, the use of platinum-based catalysts increases system cost, and catalyst degradation over time remains a concern. Another notable challenge is that the electrochemical conversion efficiency of DMFCs is generally lower than that of other fuel cells, such as solid oxide fuel cells (SOFCs).

Niche Applications and Emerging Opportunities for DMFCs

Given these characteristics, direct methanol fuel cells (DMFCs) are expected to find applications in niche markets and specific scenarios. They are currently being considered as power sources for portable electronic devices—such as mobile phones, laptops, and drones—as well as small backup power supplies for use during power outages or emergency situations. Research is also underway to evaluate their use in small vehicles, marine vessels, and lightweight electric vehicles, such as golf carts. As a clean and versatile energy source, DMFCs have considerable potential, and their application range is anticipated to expand with ongoing research and technological advancements.

Industrial Advancements in Methanol Fuel Cell Technology

Blue World Technologies is actively advancing the development and deployment of methanol fuel cell technology, with a particular focus on applications in the automotive and maritime industries. This technology utilizes methanol as a renewable fuel source, directly converting it into electricity through high-temperature polymer electrolyte membrane (PEM) fuel cells. Consequently, it offers a cleaner alternative to conventional combustion engines. Blue World Technologies has initiated series production of methanol fuel cells at its facility in Aalborg, Denmark, which is set to become the world's largest methanol fuel cell production plant. This expansion aims to support clean energy solutions that address a wide range of energy demands, including those in the heavy transport sector.

A key advantage of Blue World Technologies' methanol fuel cells is their high energy efficiency, which ranges from 40% to 60% depending on the specific application. This efficiency is further supported by a system design that enables integration into existing infrastructure with minimal modifications. The technology is designed to be cost-effective, leveraging existing liquid fuel distribution networks, thereby reducing the need for extensive new infrastructure investment. In addition, Blue World Technologies is exploring the application of methanol fuel cells in stationary power generation and other heavy-duty machinery, illustrating the versatility of this technology.

Environmental and Performance Benefits

An important environmental benefit of methanol fuel cells is their potential for carbon-neutral operation when methanol is produced from renewable resources, enabling a pathway toward net-zero carbon emissions. Roland Gumpert has developed an 850 horsepower sports car powered by a methanol fuel cell [13]. This vehicle can be fully refueled in just 3 min, has a driving range of 850 km, and can accelerate from 0 to 100 km/h in approximately 2 s.

In addition to Blue World Technologies, other companies such as Karma are also actively investigating the potential of methanol fuel cell technology. Karma's development efforts highlight the advantages of methanol as a fuel, including significantly shorter refueling times compared to battery electric vehicles, as well as the ease of storage and distribution that parallels traditional liquid fuels.

These ongoing developments and commercial initiatives underscore the potential of methanol fuel cell technology as a clean, efficient, and practical energy conversion solution for a broad range of applications.

6.5 Dimethyl Ether as a Methanol Derivative

Dimethyl ether (DME) is a derivative of methanol that has garnered significant attention as a clean fuel alternative [14]. Its chemical formula is CH_3OCH_3. DME is a colorless gaseous compound with a boiling point of −24.8 °C, making it easily liquefiable and highly volatile. It can be synthesized through the dehydration of methanol according to the following reaction:

$$2CH_3OH \rightarrow CH_3OCH_3 + H_2O \tag{6.14}$$

This synthesis is typically carried out under conditions of 1–2 MPa and 250–300 °C, utilizing solid acid catalysts such as alumina (Al_2O_3) and zeolites.

Combustion Properties and Applications

DME possesses a high cetane number (greater than 55), giving it combustion performance comparable to diesel fuel. Its high oxygen content results in soot-free combustion, thereby presenting DME as an environmentally friendly alternative to conventional compression-ignition diesel fuels. However, adapting existing diesel engines for DME use requires modifications to the fuel supply systems. DME's combustion characteristics differ from those of methanol: while methanol's high-octane number makes it suitable for spark ignition engines (e.g., gasoline engines), DME's low octane number leads to knocking in such engines, rendering it generally unsuitable for conventional spark ignition applications. Nonetheless, some research is exploring the use of DME-gasoline blends.

Beyond transportation fuels, DME can also serve as a substitute for liquefied petroleum gas (LPG) in household cooking and heating applications, often blended with propane. The ability to leverage existing LPG infrastructure allows for minimized investment in new facilities. DME's low boiling point enables it to be liquefied at relatively low pressures, similar to propane.

Among the various potential applications, one of the most promising is DME's use as a fuel for power generation. DME is well-suited for gas turbine power plants and small-scale distributed power systems. Owing to its clean combustion properties, managing exhaust emissions from power generation using DME is relatively straightforward. Another advantage of DME is its compatibility with existing infrastructure; for example, it can be supplied through LPG networks for automotive fuels, thermal power plants, and residential applications. DME's higher volumetric energy density (approximately 19.1 MJ/L compared to methanol's 15.6 MJ/L) further underscores its attractiveness. Additionally, DME is being considered for use in polymer electrolyte fuel cells, in a configuration known as a direct dimethyl ether fuel cell (DDFC), although this application is still under research and development.

Global Production and Future Outlook

As of 2024, global DME production is approximately eight million tons per year. Major producers include China, the United States, the United Kingdom/Netherlands, and South Korea. In particular, China has historically led DME production, with annual capacity exceeding ten million tons at its peak. Currently, approximately two million tons are being produced annually, primarily using coal as a feedstock, though much of the previously installed capacity remains idle. Elsewhere, natural gas is the primary feedstock for DME synthesis. In the future, there is significant potential for DME production from methanol derived from captured waste CO_2, further enhancing its sustainability profile. Beyond fuel applications, DME is also used in consumer products such as hairsprays, where it is released into the atmosphere after use.

To enable widespread adoption of DME, it is necessary to develop robust supply and storage infrastructure. While existing LPG infrastructure provides a valuable foundation, dedicated facilities may be needed to accommodate large-scale DME usage. Reducing production costs through economies of scale and advances in synthesis technology will also be crucial. Furthermore, government policies, subsidies, and regulatory support are likely to play pivotal roles in facilitating DME's expansion as a low-emission fuel alternative. Given its versatility and environmental benefits, DME is expected to play an increasingly significant role in the methanol economy as technological and infrastructure developments progress.

Industrial Initiatives and Environmental Benefits

Volvo Trucks is at the forefront of developing and field-testing vehicles powered by DME in North America. The company highlights DME's high cetane number and combustion performance, which is equivalent to diesel fuel, along with the practical advantage that DME can be stored and handled like propane, eliminating the need for high-pressure tanks. As part of its "Blue Power" strategy, Volvo is integrating DME alongside compressed natural gas (CNG) and liquefied natural gas (LNG) to offer a comprehensive portfolio of alternative fuels. Volvo's DME technology is incorporated into its trucks in a way that allows drivers to transition smoothly from conventional diesel trucks.

DME offers significant environmental advantages, including soot-free combustion. When produced from renewable resources such as biogas or renewable methanol, DME can dramatically reduce life-cycle CO_2 emissions. Volvo's efforts to develop DME-powered vehicles reflect a broader commitment to advancing sustainable energy use in the transportation sector and to moving beyond traditional fossil fuels.

Beyond Volvo, other companies are also actively engaged in DME research and deployment. For instance, Oberon Fuels is pioneering renewable DME production in the United States and has partnered with Volvo for vehicle testing. Oberon is involved in multiple projects aimed at establishing DME as a clean, practical fuel alternative. Similarly, FPT Industrial is exploring the application of DME in heavy-duty machinery, viewing it as a sustainable and efficient substitute for conventional fuels. These initiatives are supported by growing global momentum for decarbonizing transportation, as evidenced by increased funding and research initiatives. The U.S. Department of Energy has identified DME as a promising candidate for low-carbon fuel solutions.

Overall, these developments underscore the rising interest and investment in alternative fuels that offer both operational and environmental benefits, highlighting DME's role as a crucial component in the clean energy transition. Although not yet in widespread use today, the advent of the methanol society is expected to bring DME fuel into the spotlight.

References

1. Palo DR, Dagle RA, Holladay JD (2007) Methanol steam reforming for hydrogen production. Chem Rev 107:3992–4021
2. Kawasaki S, Wijaya WY, Watanabe H, Okazaki K (2011) Waste heat recovery system through methanol steam reforming with absorption heat pump. J Jpn Inst Energy 90:1152–1158
3. Lv Y, Liu F, Wang Y, Gong Y, Zhou J, Sui J (2024) Efficient hydrogen production through a novel methanol steam reforming system coupled with a two-stage heat pump and carbon capture: a thermodynamic and thermoeconomic study. Int J Hydorg Energy 88:858–877
4. Geely Homepage. Geely's E-Fuels Solution to a Carbon Neutral Future. https://global.geely.com/en/news/2024/geely-e-fuel-solution-carbon-neutral-future
5. Wärtsilä methanol engines leaflet. https://www.wartsila.com/docs/default-source/product-files/engines/wartsila-methanol-engines-leaflet.pdf?utm_source=engines&utm_medium=dfengines&utm_term=methanol&utm_content=brochure&utm_campaign=mp-engines-and-generating-sets-brochures
6. Laosiripojanaa N, Assabumrungrat S (2007) Catalytic steam reforming of methane, methanol, and ethanol over Ni/YSZ: the possible use of these fuels in internal reforming SOFC. J Power Source 163:943–951
7. Faheem HH, Abbas SZ, Tabish AN, Fan L, Maqbool F (2022) A review on mathematical modelling of direct internal reforming—solid oxide fuel cells. J Power Source 520:230857
8. Li H, Tian Y, Wang Z, Qie F, Li Y (2012) An all perovskite direct methanol solid oxide fuel cell with high resistance to carbon formation at the anode. RSC Adv 2:3857–3863
9. Sang J, Li Y, Yang J, Wu T, Luo X, Chi B, Guan W, Xu J, Singhal SC (2022) Power generation by flat-tube solid oxide fuel cells with enhanced internal reforming of methanol. ACS Sustain Chem Eng 10(19):6276–6288
10. Shekhawat D, Spivey JJ, Berry DA (2011) Fuel cells: technologies for fuel processing. Elsevier, Amsterdam
11. Behling NH (2013) Fuel cells: current technology challenges and future research needs. Elsevier, Amsterdam
12. Petrovic S, Hossain E (2020) Development of a novel technological readiness assessment tool for fuel cell technology. IEEE Access 8:132237–132252
13. Roland Gumpert Homepage, Methanol Fuel Cell. https://www.rolandgumpert.com/en/gumpert-aiways/all-news/year-in-review/
14. Dieterich V, Buttler A, Hanel A, Spliethoff H, Fendt S (2020) Power-to-liquid via synthesis of methanol, DME or Fischer–Tropsch-fuels: a review. Energy Environ Sci 13:3207–3252

Chapter 7
Chemical Use of Methanol

Abstract This chapter examines the role of methanol as a pivotal chemical feedstock and its potential to decarbonize the chemical industry. Methanol is already used extensively in the production of formaldehyde, acetic acid, methyl tert-butyl ether (MTBE), and many other chemicals. It also serves as a precursor for olefins and aromatics via methanol-to-olefins (MTO) and methanol-to-aromatics (MTA) processes. This chapter describes how methanol is industrially synthesized from CO and H_2 via catalytic hydrogenation of CO_2 and CO, using Cu/ZnO-based catalysts under moderate temperatures and high pressures. The equilibrium yield of methanol in these processes depends strongly on reaction temperature, pressure, and feed composition, making process optimization critical. Modern methanol synthesis plants operate with high efficiency and are increasingly designed to integrate renewable hydrogen and captured CO_2 to produce low-carbon or carbon-neutral methanol. Life-cycle assessments reveal that switching to CO_2-derived methanol can significantly reduce the carbon footprint of downstream chemical products. Through techno-economic and environmental comparisons, this chapter underscores methanol's scalability and feasibility as a cornerstone of sustainable chemical manufacturing.

Keywords Methanol chemistry · Chemical feedstock · MTO · Formaldehyde · Olefine · Methanol synthesis plant · Equilibrium yield · Cu/ZnO catalyst

7.1 The Future of the Methanol-Based Chemical Industry

When considering the future of the chemical industry, methanol emerges as one of the most promising primary substances. The key points supporting this assertion are as follows:

i. A circular CO_2–methanol society can be established.
ii. Methanol's dual role as both a fuel and a chemical feedstock provides significant advantages.

J. Nakamura, M. Ishida, *Blueprint for a Methanol Society*,
https://doi.org/10.1007/978-981-95-3465-4_7

iii. A wide range of chemical raw materials, including ethylene and propylene, can be synthesized from methanol.
iv. The chemical industry utilizing methanol as a feedstock is already in practical operation.
v. Biomass and organic waste can serve as raw materials for methanol synthesis.
vi. Methanol can be synthesized economically from CO_2 (or CO) and H_2.

Let us examine each of these points in detail.

Establishing a Circular CO_2–Methanol Society

Regarding points (i) and (ii), to address the ongoing global warming crisis driven by CO_2 emissions, it is imperative to reduce these emissions immediately and continuously for the foreseeable future. While fossil resources such as coal, oil, and natural gas are well-known sources of CO_2 emissions, they also serve as feedstocks for numerous manufactured products. Examples include plastic bottles, plastic bags, mobile phones, detergents, paints, inks, adhesives, construction materials, and automotive components—most of which can be synthesized using methanol as an alternative feedstock. Furthermore, once these products reach the end of their useful life, waste materials can be gasified to produce CO_2, H_2, and CO—feedstocks for methanol synthesis once again. In this way, carbon atoms can be continually recycled.

Additionally, when methanol is used as a fuel, the carbon atoms cycle between methanol (CH_3OH) and carbon dioxide (CO_2), illustrating a closed-loop carbon system. This circular approach, described in point (i), is further strengthened by the compatibility of methanol as both a fuel and a chemical feedstock, as noted in point (ii).

Methanol-to-Olefins and Other Chemical Pathways

Turning to points (iii) and (iv), methanol has the potential to become a central feedstock in the future chemical industry. Notably, it can be converted into ethylene and propylene—two of the most significant base chemicals by production volume. Ethylene ($CH_2 = CH_2$) and propylene ($CH_2 = CH–CH_3$) are olefins, hydrocarbons containing double bonds, and serve as key building blocks for polymers such as polyethylene, polypropylene, and polyvinyl chloride (PVC). The methanol-to-olefins (MTO) process enables the synthesis of these olefins from methanol. The MTO process, already in commercial operation in several countries—particularly China—typically uses coal as a feedstock for methanol production. Coal is gasified to produce CO and CO_2, which are then hydrogenated to form methanol. In the MTO process itself, catalysts such as SAPO or ZSM-5 facilitate the conversion of methanol into olefins.

This is a highly practical technology that can be rapidly implemented using existing petrochemical infrastructure. A technology that is broadly applicable across countries can be considered a true innovation, and methanol synthesis coupled with the MTO process certainly qualifies as such. Other important compounds can also be synthesized from methanol; these will be discussed in the next section.

Biomass, Waste Feedstocks, and Economic Viability

Finally, points (v) and (vi) highlight the economic viability of methanol synthesis. First, methanol can be synthesized from a wide variety of raw materials. Biomass and organic waste, in particular, are carbon-neutral resources that contribute to reducing net CO_2 emissions. Moreover, treating organic waste as a feedstock supports the recycling of carbon resources. Construction debris, forest residues, agricultural residues, and kitchen waste all consist of carbon, hydrogen, and oxygen. Through combustion with oxygen and subsequent treatment with steam (H_2O), these materials can be converted into syngas (a mixture of CO, CO_2, and H_2), which serves as a precursor for methanol synthesis.

7.2 Compounds and Products Manufactured from Methanol

Methanol serves as a highly versatile feedstock for synthesizing a wide range of industrially important chemicals, including olefins, formaldehyde, acetic acid, methylamines, and dimethyl carbonate (DMC). These intermediates are further processed into products such as plastics, rubbers, resins, pharmaceuticals, agrochemicals, solvents, and battery materials. As shown in Fig. 7.1, global methanol production increased from 73 million tons in 2014 to 110 million tons in 2024 [1]. The major contributor to this growth is the methanol-to-olefins (MTO) process, particularly in China, where ethylene and propylene production via MTO rose from nine million tons in 2014 to 34 million tons in 2024—accounting for approximately 67% of the increase.

Table 7.1 summarizes the major compounds derived from methanol. Below, we describe key production routes and applications:

i. **MTO (Methanol-to-Olefins) Process:** Methanol is thermally converted into light olefins—primarily ethylene and propylene—using shape-selective catalysts such as SAPO-34 and ZSM-5. These olefins are essential for manufacturing synthetic resins like polyethylene (PE) and polypropylene (PP), which are widely used in packaging, textiles, consumer goods, and automotive components.
ii. **MTG (Methanol-to-Gasoline) Process:** The MTG process converts methanol into high-octane gasoline via dehydration to dimethyl ether, followed by cata-

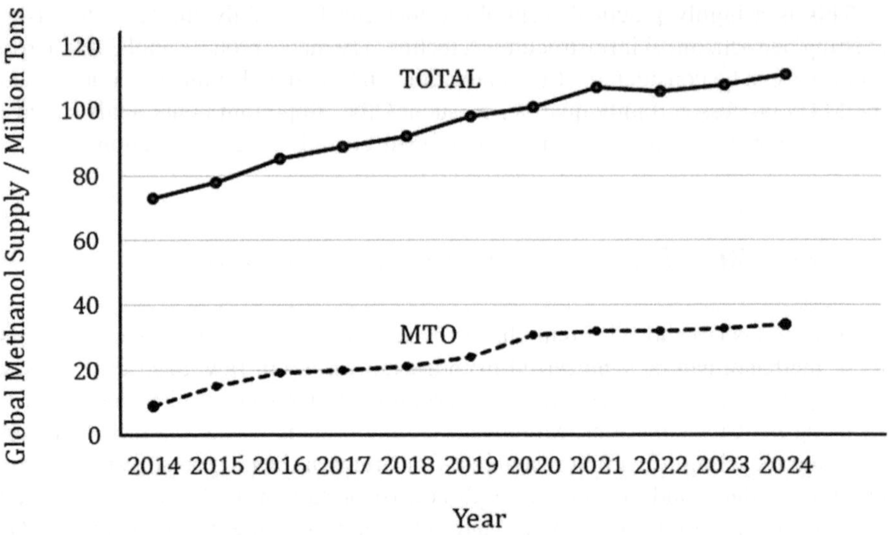

Fig. 7.1 Annual changes in global methanol supply and consumption in the MTO process [1]

Table 7.1 Various chemical raw materials manufactured using methanol

Condensed formula	Compound name	Derivative
C_2H_4	Ethylene	Polyethylene
C_3H_6	Propylene	Polypropylene
HCHO	Formaldehyde	Urea-formaldehyde resin, phenol-formaldehyde resin, melamine-formaldehyde resin
CH_3OCH_3	Dimethyl ether (DME)	Methyl acetate, acetic acid, butadiene, fuel
C_6H_6, $C_6H_5CH_3$, $C_6H_4(CH_3)_2$	Benzene, toluene, xylene (BTX)	Polystyrene, phenol, PET
CH_3COOH	Acetic acid	Vinyl acetate, acetic anhydride, ethyl acetate
$CH_2C(CH_3)COOCH_3$	Methyl methacrylate (MMA)	Polymethyl methacrylate (PMMA)
CH_3NH_2	Methylamines	Dimethylformamide (DMF), chloromethyl compounds
$(CH_3O)_2O$	Dimethyl carbonate (DMC)	Polycarbonate (PC), methyl carbamate, ethyl methyl carbonate (EMC)
$(CH_3)_3COCH_3$	Methyl tertiary-butyl ether (MTBE)	Fuel additive

lytic conversion over zeolites. Using ZSM-5, light olefins are transformed into gasoline-range hydrocarbons. MTG enables synthetic fuel production from methanol derived from natural gas, coal, or CO_2 and hydrogen. These hydrocarbons are also used as solvents and as feedstocks for synthetic resins and other chemicals.

iii. **MTA(Methanol-to-Aromatics) Process:** Benzene, toluene, and xylene—collectively known as BTX aromatics—can be synthesized from methanol through methanol-to-aromatics (MTA) or methanol-to-hydrocarbons (MTH) processes using modified zeolite catalysts such as ZSM-5. These aromatics are fundamental building blocks in the petrochemical industry. Benzene is used to produce styrene (for polystyrene), phenol (for epoxy resins), and nylon precursors. Toluene serves as a solvent and a feedstock for toluene diisocyanate (TDI), used in polyurethane foams. Xylenes, particularly para-xylene, are oxidized to terephthalic acid for the manufacture of polyethylene terephthalate (PET) used in plastic bottles and textile fibers.

This methanol-based route provides an alternative to conventional naphtha cracking and can contribute to decarbonizing aromatic production when methanol is derived from CO_2 and green hydrogen.

iv. **Synthesis of Formaldehyde:** Formaldehyde is produced via the catalytic partial oxidation of methanol using silver or iron-molybdenum catalysts. It is a key intermediate for resins such as urea-formaldehyde (UF), phenol-formaldehyde (PF), and melamine-formaldehyde (MF), which are used in wood products, coatings, and adhesives.

v. **Synthesis of Acetic acid:** Acetic acid is manufactured by carbonylation of methanol with carbon monoxide in the presence of rhodium or iridium catalysts (Monsanto or Cativa process). It is used in the production of vinyl acetate monomer (VAM), acetic anhydride, and esters for solvents, plastics, and textile fibers.

vi. **Synthesis of Methylamine:** Methylamines—monomethylamine (MMA), dimethylamine (DMA), and trimethylamine (TMA)—are synthesized by the reaction of methanol with ammonia. These are important intermediates for the production of pesticides (e.g., herbicides and fungicides), pharmaceuticals, surfactants, and solvents.

vii. **Synthesis of Dimethyl Carbonate (DMC):** DMC is produced via the oxidative carbonylation of methanol using CO or CO_2. DMC is a non-toxic, biodegradable solvent used in paints and coatings and also serves as a precursor for polycarbonate plastics and electrolytes in lithium-ion batteries.

As described above, various compounds can be synthesized from methanol. If ethanol derived from biomass is also utilized, it becomes feasible to construct a chemical industry system centered on both methanol and ethanol. This integrated approach is expected to contribute significantly to the establishment of a carbon recycling society. Such a transformation not only promotes the development of a future recycling-oriented economy but also addresses the pressing global challenge of reducing CO_2 emissions.

7.3 Current Methods of Methanol Production

Methanol synthesis technology is already well-established, with global methanol production currently amounting to approximately 110 million tons annually. Let us begin by reviewing the present state of methanol production.

Methanol is synthesized from carbon monoxide (CO) and hydrogen (H_2), as shown in Reaction 7.1. The mixture of CO and H_2 is referred to as synthesis gas, as it can be used to produce methane, methanol, and various hydrocarbons. In most countries, synthesis gas is generated via the steam reforming of methane (CH_4), represented by Reaction 7.2. Consequently, natural gas, which primarily consists of methane, is the principal feedstock. Additionally, synthesis gas can be produced by coal gasification, and in China, methanol production from coal-based synthesis gas is widely practiced.

$$CO + 2H_2 \rightarrow CH_3OH \ (\mathrm{Cu/ZnO\ catalyst}) \tag{7.1}$$

$$CH_4 + H_2O \rightarrow CO + 3H_2 \ (\mathrm{Ni\ catalyst}) \tag{7.2}$$

Role of CO_2 and Catalyst Stabilization

During methanol synthesis from synthesis gas, carbon dioxide (CO_2) is typically added to the feedstock to mitigate the excessive reduction of the Cu/ZnO catalyst by the strongly reducing synthesis gas. Both CO and H_2 are potent reducing agents, and in their presence, the oxygen from ZnO can be stripped away, leading to catalyst degradation via sintering. The inclusion of CO_2 or H_2O, as oxidizing agents, helps maintain Cu and ZnO in a moderately oxidized state, thereby stabilizing catalyst activity.

Extensive research has been conducted on the Cu/ZnO catalytic system, and it has been established that methanol is more readily synthesized from CO_2 than from CO. In other words, the reaction rate for methanol formation from CO_2 is higher than from CO. Consequently, methanol can also be synthesized directly from CO_2 and H_2 via Reaction 7.3, using the same Cu/ZnO-based catalyst. A further advantage is the affordability and widespread availability of these catalyst materials.

$$CO_2 + 3H_2 \rightarrow CH_3OH + H_2O \ (\mathrm{Cu/ZnO\ catalyst}) \tag{7.3}$$

Despite this, industrial methanol production does not currently rely on CO_2 as the primary feedstock. This is because existing methanol production facilities are closely integrated with steam reforming of methane.

Reaction Conditions and Performance Metrics

Tables 7.2 and 7.3 summarize the reaction conditions and yields for methanol synthesis from CO and CO_2, respectively, based on experimental data from industrial and research facilities [2]. In all cases, Cu/ZnO catalysts are employed, and the reaction conditions typically involve temperatures around 250 °C and pressures of

Table 7.2 Commercial methanol synthesis from CO and H_2 [2]

Licensor	T (°C)	P (bar)	GHSV (×10³ h⁻¹)	STY (kg L_{cat}^{-1} h^{-1})
Süd Chemie	250	50	10	1.03–1.10
BASF	230	50	10	1.43
ICI	240	51	9.6	0.43
Shell	250	53	11.5	0.69
Shell	300	53	10.9	1.01
Casale	250	100	12.5	1
MGC	264	100	10	2.22
Topsoe (MK-121)	200–310	39–122	n.a.	n.a.
Topsoe (MK-101)	221	67	n.a.	1.02
ICI	260	41	6–7	0.26
ICI	260	41	10	0.38
ICI	260	81	10	0.77
Metallgesellschaft	250	41	9.8	1.42
Metallgesellschaft	250	101	9.8	2.28
NIRE/RITE	250	50	10	1.7

Table 7.3 Commercial methanol synthesis from CO_2 and H_2 [2]

Licensor	T (°C)	P (bar)	GHSV (×10³ h⁻¹)	STY (kg L_{cat}^{-1} h^{-1})
Süd Chemie	250	80	10.5	0.6
NIRE/RITE	250	50	18	0.75
NIRE/RITE	250	30–70	10	0.4–0.8
NIRE/RITE	250	50	10	0.76
JM Katalco 51–8	240	69–97	3.3–8.3	0.06–0.24
Süd Chemie (C79–05-GL)	260	80	8.1	0.66
Commercial	250	30	7.9	0.6

approximately 50 atm. Under these conditions, there is little difference in the hydrogenation reactions of CO and CO_2.

The gas hourly space velocity (GHSV), expressed in h^{-1}, indicates the volumetric flow rate of reaction gas per unit catalyst volume, converted to standard conditions. The reciprocal of GHSV represents the residence time (or reaction time) of the reaction gas within the catalyst bed. The space time yield (STY), expressed in $kg \cdot L_{cat}^{-1} \cdot h^{-1}$, measures the mass of product formed per hour per unit catalyst volume. Alternatively, STY can be expressed in terms of the number of moles of product or the mass of catalyst used.

Analysis of the data in Tables 7.2 and 7.3 reveals that there is minimal difference in reaction rates between CO and CO_2 hydrogenation at GHSV = 10,000 h^{-1}. The residence time is 3600/10,000 = 0.36 s. In fact, even in the case of CO hydrogenation, CO_2 and H_2O are contained in the reaction gas, and it is believed that CO is converted to methanol via CO_2. However, STY values are different for CO hydrogenation, 1–2 $kg \cdot L_{cat}^{-1} \cdot h^{-1}$, and for CO_2 hydrogenation, 0.5–1 $kg \cdot L_{cat}^{-1} \cdot h^{-1}$, reflecting the higher equilibrium conversion of CO (as shown in Fig. 7.2) compared to CO_2 (Fig. 7.3).

Fig. 7.2 Photograph of the CO_2 to methanol plant operating with CRI's ETL technology™ located at the Shenghong Petrochemical Industrial Park in Jiangsu, China (Sailboat Project), Home page of Carbon Recycling International, https://carbonrecycling.com/about/news/carbon-recycling-international-cri-and-jiangsu-sailboat-start-up-worlds-most-efficient-co2-to-methanol-plant Photo courtesy of the Carbon Recycling International (CRI)

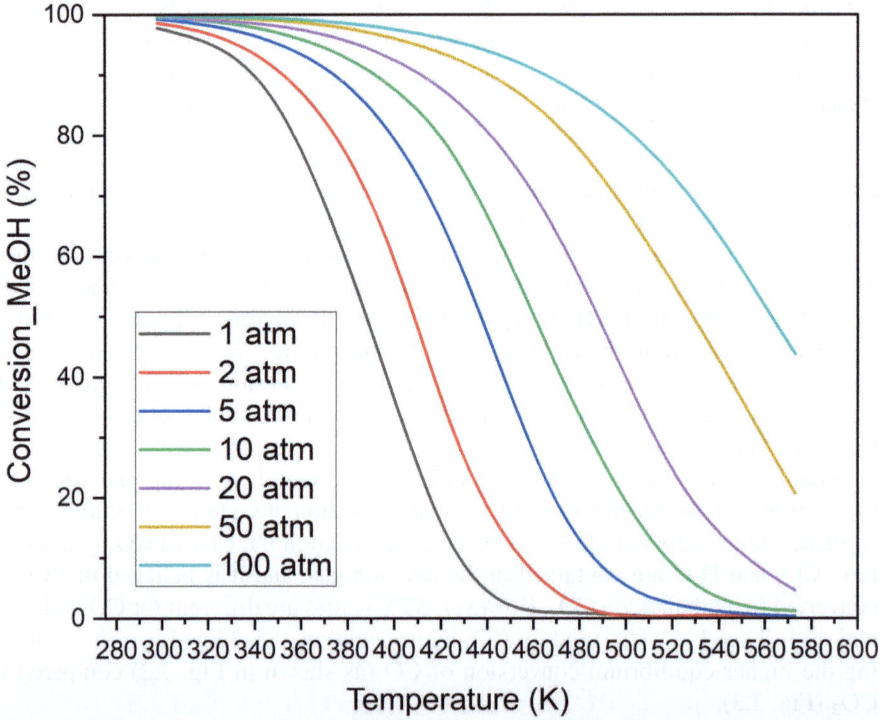

Fig. 7.3 Equilibrium yield of methanol synthesis from CO. $P(CO)/P(H_2) = 1/2$

Implications for Industrial Energy Demand

Understanding these reaction rates provides insight into the energy requirements for methanol and ammonia synthesis. A low absolute reaction rate necessitates greater energy input to maintain production rates. Ammonia synthesis, for instance, is recognized as an energy-intensive industrial process, accounting for 1–2% of global energy consumption. This substantial energy demand makes ammonia less suitable as an energy carrier.

The GHSV for ammonia synthesis ranges from 10,000 to 100,000 h^{-1}, with STY values of 1–2 kg · $L_{-}cat^{-1}$ · h^{-1}—comparable to those for methanol synthesis. However, ammonia synthesis operates under significantly higher temperatures and pressures: 400–500 °C and 150–350 atm, compared to 250 °C and 50–100 atm for methanol synthesis. Increasing reaction temperature and pressure is a standard approach to accelerate sluggish chemical reactions, but this also requires substantial energy input.

The rate constant (k) is a useful parameter for comparing reaction rates:

$$k = A\cdot\exp\left(-\frac{E}{RT}\right) \tag{7.4}$$

where A is the pre-exponential factor, E is the activation energy, R is the gas constant, and T is the reaction temperature (in Kelvin).

Assuming similar activation energies and pre-exponential factors for methanol and ammonia synthesis (E = 60–100 kJ/mol), the effect of temperature on reaction rates can be estimated. For E = 60 kJ/mol and E = 100 kJ/mol, the relative increase in the rate constant when raising the temperature from 523 K (methanol synthesis) to 673 K (ammonia synthesis) is:

For E = 60 kJ/mol:

$$\frac{k(673\text{K})}{k(523\text{K})} = \exp\left\{\frac{60{,}000}{8.31}\left(\frac{1}{523}-\frac{1}{673}\right)\right\} \approx 21.7$$

For E = 100 kJ/mol:

$$\frac{k(673\text{K})}{k(523\text{K})} = \exp\left\{\frac{100{,}000}{8.31}\left(\frac{1}{523}-\frac{1}{673}\right)\right\} \approx 168$$

These calculations show that increasing the temperature from 523 K to 673 K can enhance the reaction rate by 20–170 times. This substantial difference illustrates why ammonia synthesis requires such high temperatures and pressures: the inherent reaction rate is significantly lower. It also suggests that, compared to methanol synthesis, the pre-exponential factor A in ammonia synthesis may be one to two orders of magnitude smaller.

7.4 Future Methods of Methanol Production

As we move toward a carbon-neutral future, the following methods of producing methanol are expected in the short, medium, and long term.

i. Methanol synthesis using **natural gas** as a source of hydrogen and carbon.
ii. Methanol synthesis using H_2, CO, and CO_2 derived from **biomass**.
iii. Methanol synthesis using **waste plastic** as carbon source.
iv. Methanol synthesis from **green hydrogen** produced using renewable energy and **CO_2 captured** from industrial sources (Cu/ZnO catalyst).
v. **Direct electrochemical methanol synthesis** from CO_2 and H_2O.

Among these methods, (i) to (iii) are considered **effective short-term strategies**. Method (i), which relies on natural gas—a fossil fuel—serves as an initial step for establishing a methanol supply chain and can be implemented immediately. Methods (ii) and (iii), which utilize biomass and waste plastics, respectively, may not immediately enable large-scale methanol production but can contribute meaningfully to CO_2 emission reduction. Using waste plastic as a carbon source aligns well with waste management and recycling initiatives, making it one of the most future-oriented approaches. Method (iv) is a central hydrogen production method for the hydrogen society, utilizing renewable energy, and is attracting public attention. Although the technology is established, high costs prevent it from being adopted for large-scale production immediately. Method (iv) is thus considered a promising medium-term goal. Finally, method (v), the direct electrochemical synthesis of methanol from CO_2 and water, remains at the research stage. Nevertheless, it holds significant potential for future implementation and warrants continued research and development focus.

Table 7.4 summarizes some reports on methanol production costs. In the process of producing methanol from CO_2 and renewable energy, two key stages dominate the overall cost: CO_2 capture and H_2 production. In particular, hydrogen production represents a substantial portion of the cost. For instance, data from 2019 to 2022 indicate that the cost of producing methanol using green hydrogen is approximately USD 1000 per ton of methanol. In contrast, when using hydrogen derived from natural gas, the cost is about USD 350 per ton. This USD 650 per ton difference reflects the cost of hydrogen production via renewable energy sources. Notably, it has been reported that hydrogen production accounts for approximately 70% of the total methanol production cost [3]. This high proportion actually highlights the relative cost-effectiveness of methanol production itself, with hydrogen generation (via water electrolysis) being the primary expense. Most of the electricity generated from renewable sources is dedicated to water electrolysis for hydrogen production. Currently, in conventional methanol production, hydrogen is obtained from the steam reforming of methane. However, advances in hydrogen production technology and the scaling-up of green hydrogen production are expected to drive costs down in the future.

Table 7.4 Methanol production cost [4–7]

Source	From renewable energy	From natural gas, oil, coal	From biomass
National Renewable Energy Laboratory (NREL) (2020)	800 USD	310–420 USD	543 USD
International renewable energy agency (IRENA) (2021)	800–1200 EUR	100–250 USD	N/A
Frontiers in Energy Research (2019)	868 EUR	N/A	N/A
Sustainable energy fuels, 2, 1244 (2018)	548 USD	N/A	N/A
S&P Global (2022)	N/A	344–395 USD	N/A

As discussed in Part III, a desirable long-term goal is to develop technologies for direct electrochemical synthesis of methanol from CO_2 and water, bypassing the separate water electrolysis stage.

Challenges in CO_2 Capture and Utilization

The cost of CO_2 capture itself is also significant, estimated to represent around 10% of the total methanol production cost and cannot be ignored. Lowering this cost is another important area of development. CO_2 is typically captured from industrial exhaust gases or directly from the atmosphere through technologies such as absorption, adsorption, and membrane separation. These technologies are the focus of active research and development efforts.

Once CO_2 is captured and hydrogen is produced, methanol is synthesized under high-temperature and high-pressure conditions in the presence of a catalyst. A copper-zinc-alumina (Cu-ZnO-Al_2O_3) catalyst is typically used for this purpose, enabling efficient methanol production.

Current Industrial Implementation

Recently, as shown in Table 7.5, plants synthesizing methanol from CO_2 have been rapidly established worldwide. **Carbon Recycling International (CRI)** is recognized as a pioneer in industrial CO_2-to-methanol production. CRI and its partners currently operate facilities with a combined capacity of approximately 214,000 tons of methanol per year, utilizing about 310,000 tons of captured CO_2 (as of 2025). In 2012, CRI launched the George Olah plant in Iceland, which produces methanol using CO_2 from geothermal power plants and hydrogen generated via renewable electricity.

Table 7.5 Methanol plant using CO_2 (As of 2025)

Plant name	Country	CO_2 usage (tons/year)	Methanol production (tons/year)	Hydrogen production method	CO_2 source
George Olah plant (2012~)	Iceland	5500	4000	Electrolysis of water	Geothermal power plants
Shunli plant (with CRI, 2022~)	China (Anyang)	160,000	110,000	Produced from coke oven gas (COG)	Coke oven gas (COG)
Sailboat project (with CRI, 2023~)	China (Lianyung-ang)	150,000	100,000	Chemical by-products	From ethylene oxide production facility
Finnfjord (with CRI)	Norway		100,000 (planed)	Electrolysis of water using renewable energy	From the emissions of the Finnfjord ferrosilicon plant
Tianying (with CRI)	China		170,000 (planed)	Electrolysis of water using renewable energy	CO_2 from biomass combustion
Fairway methanol LLC (2024~)	USA	180,000 (planed)	162,000 (planed)	Blue hydrogen (2024)	CO/H_2 production facility (Linde)
Kassø e-methanol (European Energy, 2025~)	Denmark	60,000	42,000	Electrolysis of water using renewable energy	Biomass and waste

Since then, CRI's Emissions-to-Liquids (ETL) technology has been adopted internationally, most notably in large-scale commercial plants in Anyang and Lianyungang, China. The Shunli plant produces about 110,000 tons of methanol annually from CO_2 and H_2 recovered from coke oven gas in steelmaking. Although its direct contribution to carbon neutrality is limited, it plays an important role in establishing the methanol value chain. The Sailboat project in Lianyungang shown in Fig. 7.2, also based on CRI's ETL technology, produces 100,000 tons of methanol per year using CO_2 and H_2 obtained as by-products from on-site processes, such as waste gas from an ethylene oxide unit within the same petrochemical complex.

In Norway, the Finnfjord project plans to use CO_2 from the exhaust gas of a ferrosilicon plant and green hydrogen produced through electrolysis powered by renewable energy, with a targeted capacity of 100,000 tons of methanol per year. Similarly, the Tianying project in Liaoyuan, China, is constructing a facility that will synthesize methanol from biomass-derived CO_2 and renewable hydrogen, with an initial capacity of 170,000 tons per year.

Beyond China and Iceland, CO_2-based methanol production has already begun in the United States—where the Fairway Methanol JV commenced operations in

2024—and in Denmark, with the Kassø e-Methanol facility inaugurated in 2025. As of 2025, further projects are planned in Germany, Belgium, Sweden, and Canada. While these developments represent a promising trend, the scale of methanol production from CO_2 must be significantly accelerated to address the urgency of climate change.

Prospects for Direct Electrochemical Methanol Synthesis

Finally, let us discuss the electrochemical synthesis of methanol from CO_2 and H_2O. Throughout this section, we have emphasized that hydrogen production represents a significant portion of the total cost of methanol production. Ideally, methanol could be synthesized directly from CO_2 and H_2O, bypassing the intermediate step of hydrogen production altogether. Recently, an iron porphyrinoid catalyst capable of direct electrochemical synthesis of methanol from CO_2 has been reported [8]. However, it will take considerable time and effort to put it to practical use. Nevertheless, an intermediate product with substantial promise is **formic acid**. Electrochemical synthesis from CO_2 and H_2O can readily produce formic acid. Subsequently, methanol can be thermally synthesized from formic acid using a **Cu/ZnO catalyst**. This two-step pathway—formic acid generation followed by thermal conversion to methanol—may provide a feasible approach for methanol production with a lower energy input in the future.

7.5 Methanol Synthesis Catalyst

Equilibrium and Kinetics

Methanol can be synthesized relatively easily from CO_2; however, the low conversion rate under industrial conditions presents a challenge. Typically, methanol synthesis is carried out at a reaction temperature of approximately **523 K** and a pressure of about **5 MPa** (50 atmospheres). Under these conditions, due to equilibrium and kinetic limitations, no more than 20% of the CO_2 can be converted to methanol in a single pass. However, from an equilibrium perspective, lowering the reaction temperature could theoretically increase the methanol yield to over 50%.

To understand this in detail, let us examine the **equilibrium and kinetics** of methanol synthesis. First, consider the equilibrium of methanol synthesis from both CO and CO_2, as well as the associated **water-gas shift reaction** that can occur as a side reaction. The standard enthalpy changes at 25 °C for methanol synthesis from CO and CO_2 are as follows:

$$CO + 2H_2 \rightleftharpoons CH_3OH(g) \quad \Delta_r H^\circ = -90.7\,kJ/mol \tag{7.5}$$

$$CO_2 + 3H_2 \rightleftharpoons CH_3OH(g) + H_2O(g) \quad \Delta_r H^\circ = -49.5\,kJ/mol \tag{7.6}$$

These enthalpy values indicate that methanol synthesis from CO is more exothermic than from CO_2. Consequently, from a thermodynamic perspective, methanol production from CO is more favorable. At equivalent temperature and pressure conditions, the equilibrium yield of methanol from CO hydrogenation is significantly higher than that from CO_2 hydrogenation.

Figures 7.3 and 7.4 illustrate the equilibrium methanol yields for CO and CO_2 hydrogenation, respectively, calculated by us. These agree with literature data [9]. For example, at the typical industrial reaction temperature of **523 K**, the equilibrium yield for methanol from CO_2 hydrogenation is approximately **20%** at 5 MPa and **30%** at 10 MPa. In contrast, for CO hydrogenation, the equilibrium yield reaches **90%** at 5 MPa and **95%** at 10 MPa. These values represent **one-pass yields**, i.e., the percentage of CO or CO_2 converted to methanol in one pass through the reactor, before any gas recycling. When the one-pass yield is low, unreacted gases are typically recycled to improve overall process efficiency.

As shown in these figures, reducing the reaction temperature increases the equilibrium yield. However, this comes at the cost of a significant decrease in the reaction rate. This creates a **trade-off**: lower temperatures are advantageous for equilibrium conversion, while higher temperatures favor reaction kinetics. As a result, the optimal reaction temperature represents a balance between these competing effects. For CO_2 hydrogenation, it is particularly important to either operate at lower temperatures to improve equilibrium yields or to remove the water formed during the reaction to drive the equilibrium toward methanol formation.

In analyzing the equilibrium behavior of CO_2 hydrogenation, the **(reverse) water-gas shift reaction** must always be considered:

$$CO_2 + H_2 \rightleftharpoons CO + H_2O(g) \quad \Delta_r H^\circ = 41.2\,kJ/mol \tag{7.7}$$

The equilibrium yields of CO in CO_2 hydrogenation (CO_2:H_2 = 1:3) are shown in Fig. 7.5. As discussed later (Sect. 11.3), it is well established that, within the Cu-catalyst system, the rate of CO production via the reverse water-gas shift reaction is orders of magnitude faster than the rate of methanol synthesis from CO or CO_2. Also, the rate of methanol synthesis from CO_2 is greater than the rate of methanol synthesis from CO. Industrially, methanol is usually produced by CO hydrogenation, but CO_2 and H_2O are intentionally introduced. This has two major implications: first, to prevent catalyst sintering caused by the highly reducing CO/H_2 reaction mixture, and second, to efficiently synthesize methanol by converting CO to CO_2. Accordingly, a detailed understanding of the reverse water-gas shift reaction is essential for methanol synthesis process design and catalyst development.

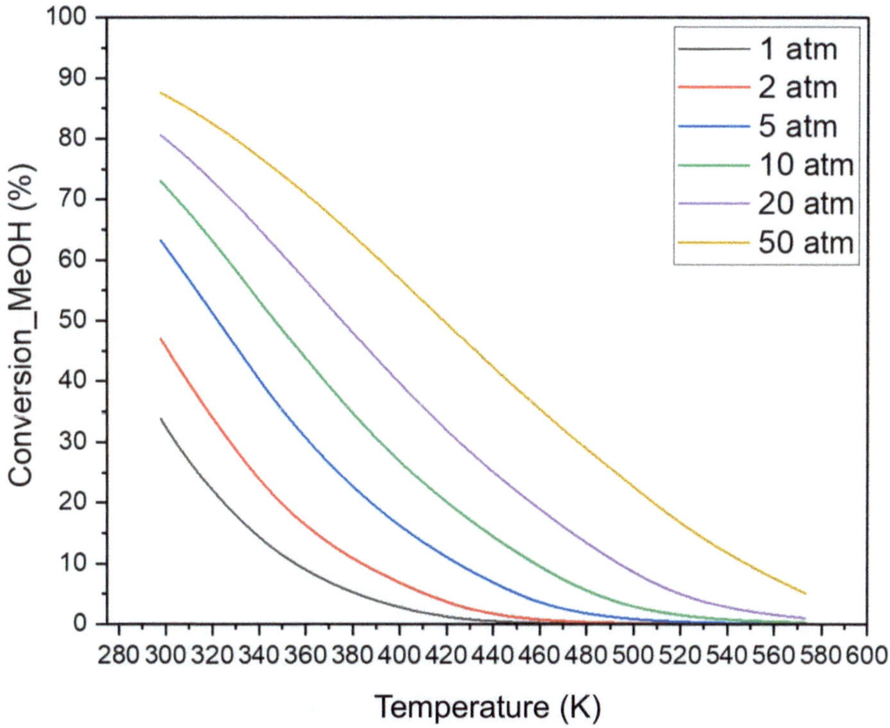

Fig. 7.4 Equilibrium yield of methanol synthesis from CO_2 (considering water-gas shift reaction). $P(CO_2)/P(H_2) = 1/3$

Catalyst Design and Function

A catalyst is essential for methanol synthesis. While a catalyst promotes the reaction by increasing its rate, it does not alter the reaction equilibrium. Consequently, catalysts do not appear in the equilibrium equations discussed earlier; their role is solely described within the framework of reaction kinetics.

As previously mentioned, a Cu/ZnO catalyst is employed industrially for methanol synthesis. The typical Cu/Zn ratio in these catalysts ranges from 70:30 to 50:50, consisting of Cu particles and ZnO particles with sizes ranging from a few nanometers to several tens of nanometers as shown in Fig. 7.6. The active sites responsible for catalysis are primarily provided by the Cu particles. Meanwhile, the ZnO particles act as spacers, preventing sintering of the Cu particles and contributing to the formation of active sites on the Cu surface.

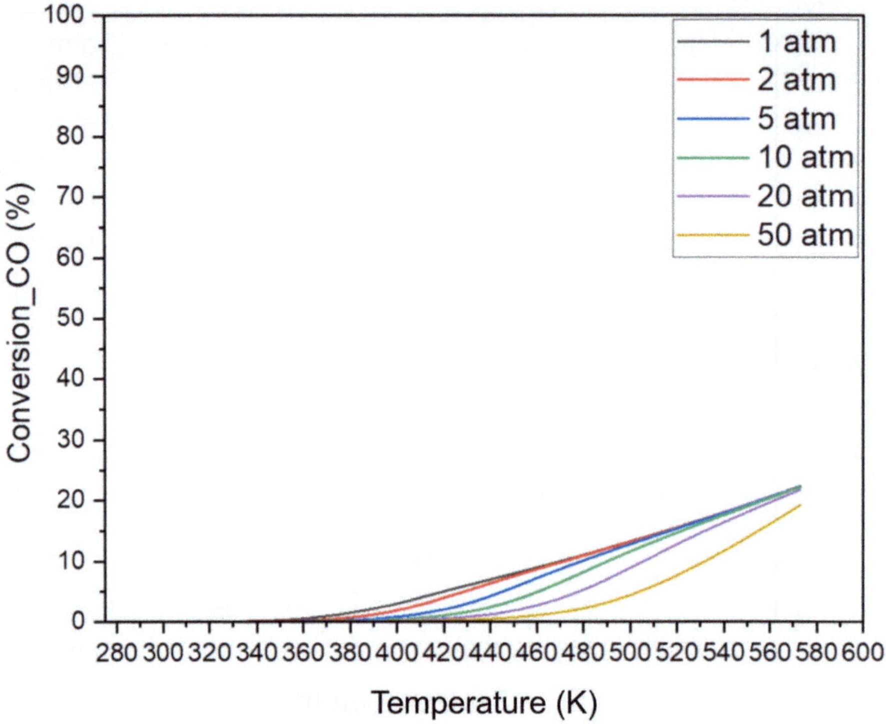

Fig. 7.5 Equilibrium yield of CO from CO_2 hydrogenation (considering methanol formation). $P(CO_2)/P(H_2) = 1/3$

As will be discussed in detail in Sect. 11.3, there is ongoing debate regarding the precise nature of these active sites, with two principal models proposed: the CuZn alloy site model and the Cu/ZnO interface model.

In industrial catalysts, Al_2O_3 is added to the Cu/ZnO catalyst to stabilize the Cu and ZnO particles during reaction conditions. Additionally, it has been reported that the incorporation of SiO_2 can help mitigate catalyst deactivation.

The Cu/ZnO/Al_2O_3 catalyst is utilized for methanol synthesis from both CO and CO_2. In industrial methanol synthesis from CO, the feed gas consists of high-pressure CO and H_2—both of which are highly reducing gases. At the typical reaction temperature of about 523 K, these conditions lead to significant catalyst reduction, causing sintering of the catalyst particles, an increase in Cu particle size, and a reduction in the effective catalyst surface area. This phenomenon is a major issue in methanol synthesis. To counteract this, CO_2 and H_2O are introduced into the reaction gas to prevent excessive catalyst reduction.

Interestingly, when a Cu/ZnO catalyst is used, methanol can be synthesized from both CO_2 and CO. However, it has been observed that the reaction rate is higher when starting from CO_2, suggesting that the reaction mechanisms and active sites for methanol synthesis differ depending on the reactant. Under actual industrial reaction conditions, methanol is thought to be produced from both CO_2 and CO simultaneously. This topic will be explored in greater detail in Sect. 11.3.

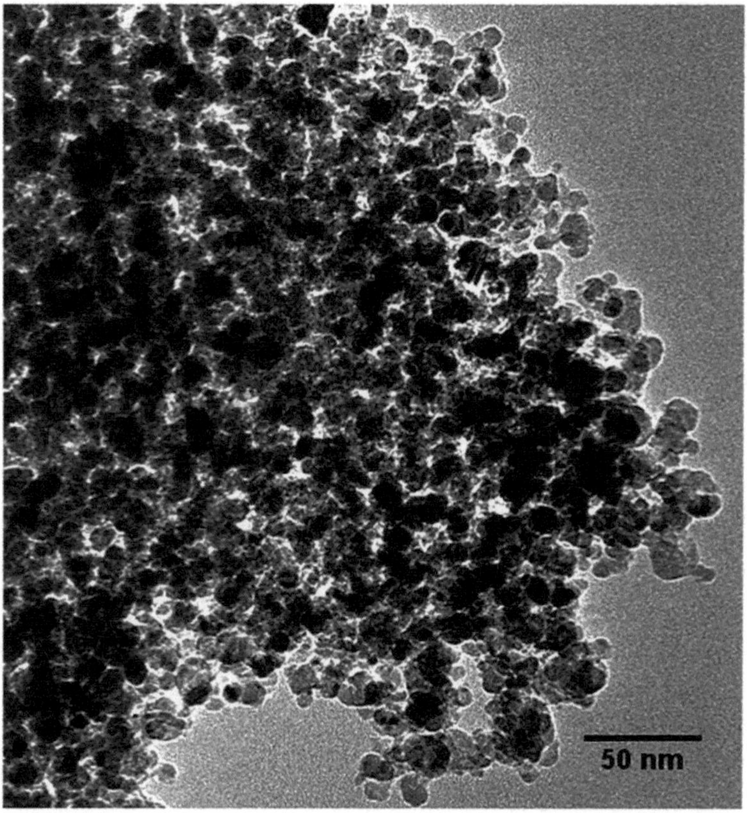

Fig. 7.6 Electron microscope image of Cu/ZnO catalyst. (Reprinted from [10] with the permission of The Royal Society of Chemistry)

Cost Considerations and Advanced Reactor Design

In addition to the widely used Cu/ZnO catalyst, it is known that a Pd/ZnO catalyst also exhibits catalytic activity for methanol synthesis. However, the cost difference is substantial: as of July 2024, palladium is priced at approximately 32,760 USD/kg, whereas copper costs around 9 USD/kg—making palladium over 3000 times more expensive. Other precious metal catalysts, such as iridium (5000 USD/kg), rhenium (1300 USD/kg), and gold (76,533 USD/kg), have also been reported to show activity. Nevertheless, considering the scale required for a methanol-based society, the enormous volume of catalyst needed renders cost a decisive factor for practical adoption. Currently, Cu/ZnO remains the only promising practical catalyst for large-scale methanol synthesis. This is visually represented in the accompanying cost bar graph. The cost of catalysts in general will also be discussed in Sect. 11.2.

Given these circumstances, research efforts have focused on increasing methanol yield through reactor design based on the Cu/ZnO catalyst. Several studies have reported attempts to enhance methanol yield by removing water, a by-product that can inhibit the reaction. One effective approach involves the incorporation of separation membranes within the reactor system [11, 12].

Figure 7.7 shows a flowchart of a reactor system that incorporates intermediate condensation units to sequentially remove methanol and water—the main reaction products [11]. This reactor is divided into three modules, with an intermediate condensation unit placed between each module. Through cooling and separation processes, the water concentration in the reactor is kept low. Under these conditions, a one-pass CO_2 yield of 53.9% and a methanol selectivity of 99.5% have been achieved. Since water can contribute to catalyst degradation, removing it not only

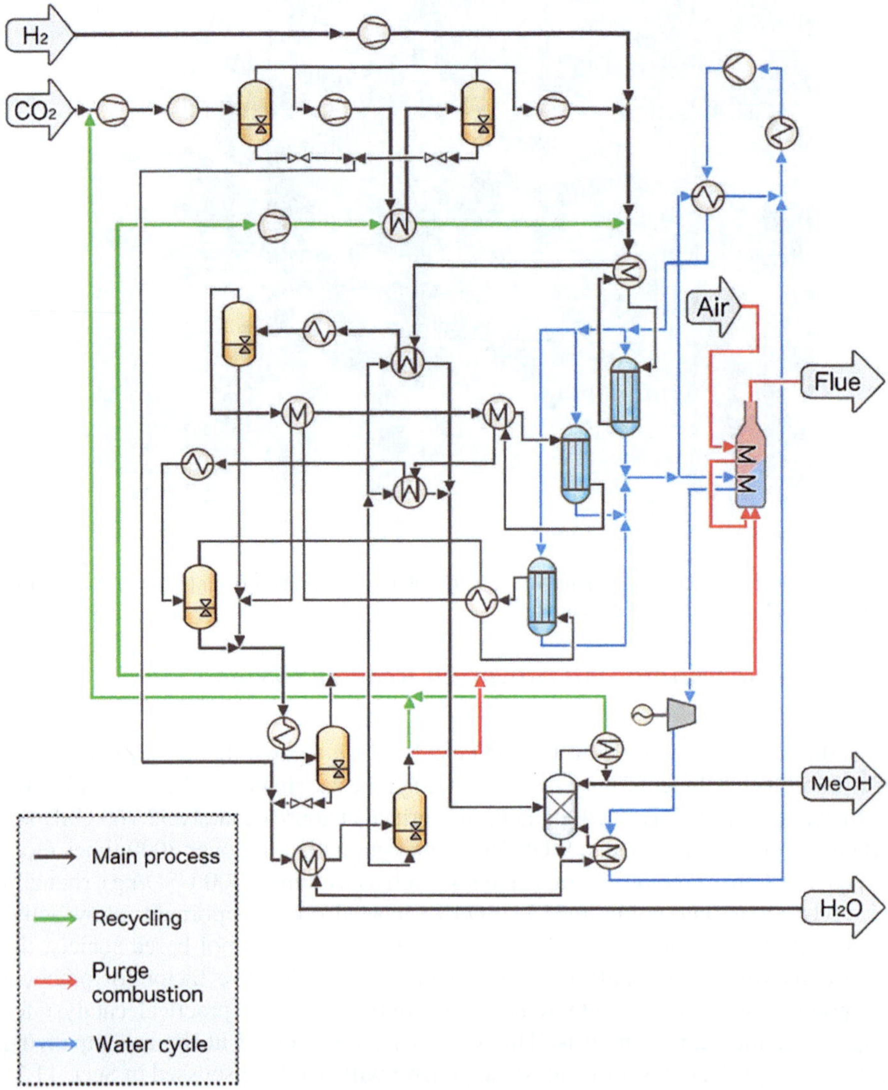

Fig. 7.7 Flowchart of methanol synthesis from CO_2. Adapted from [11] under the Creative Commons Attribution 4.0 International License

improves yield but also extends catalyst lifetime. These reactor designs belong to the class of "membrane-assisted" or "modular" reactors, which allow in situ product separation to enhance conversion and catalyst longevity.

References

1. Homepage of Methanol Institute (2024) Methanol Price, Data from MMSA. https://www.methanol.org/methanol-price-supply-demand/
2. Dieterich VV, Buttler A, Hanel A, Spliethoff H, Fendt S (2020) Power-to-liquid via synthesis of methanol, DME or Fischer–Tropsch-fuels: a review. Energy Environ Sci 13:3207–3252
3. Gonzalez-Garay A, Frei MS, Al-Qahtani A, Mondelli C, Guillen-Gosalbez G, Perez-Ramırez J (2019) Plant-to-planet analysis of CO_2-based methanol processes. Energy Environ Sci 12:3425–3436
4. Borisut P, Nuchitprasittichai A (2019) Methanol production via CO_2 hydrogenation: sensitivity analysis and simulation—based optimization. Front Energy Res 7–81:1–10
5. Hank C, Gelpke S, Schnabl A, White RJ, Full J, Wiebe N, Smolinka T, Schaadt A, Henning H-M, Heblinga C (2018) Economics & carbon dioxide avoidance cost of methanol production based on renewable hydrogen and recycled carbon dioxide—power-to methanol. Sus Energy Fuels 2:1244–1261
6. Irena and Methanol Institute (2021) Innovation outlook: renewable methanol. International Renewable Energy Agency, Abu Dhabi. https://www.irena.org/Publications/2021/Jan/Innovation-Outlook-Renewable-Methanol
7. Harris K, Grim R, Tao L (2020) A comparative techno-economic analysis of renewable methanol synthesis pathways from biomass and CO_2. National Renewable Energy Laboratory, Golden, CO. NREL/CP-5100-78547
8. Saha P, Amanullah SK, Barman S, Dey A (2025) Electrochemical reduction of CO_2 to CH_3OH catalyzed by an iron Porphyrinoid. J Am Chem Soc 147:1497–1507
9. Ohyama S (1995) Evaluation of catalytic hydrogenation of carbon dioxide to methanol. J Jpn Inst Energy 74:137–148
10. Behrens M, Lolli G, Muratova N, Kasatkin I, Havecker M, Naumann R, d'Alnoncourt NR, Storcheva O, Kohler K, Muhlerd M, Schlogl R (2013) The effect of Al-doping on ZnO nanoparticles applied as catalyst support. Phys Chem Phys Chem 15:1374–1381
11. Campos BLO, John K, Beeskow P, Delgado KH, Dahmen SPN, Sauer J (2022) Detailed process and techno-economic analysis of methanol synthesis from H_2 and CO_2 with intermediate condensation steps. Processes 10:1535. 1-24. CC BY 4.0
12. Seshimo M, Liu B, Lee HR, Yogo K, Yamaguchi Y, Shigaki N, Mogi Y, Kita H, Nakao S (2021) Membrane reactor for methanol synthesis using Si-rich LTA zeolite membrane. Membranes 11–505:1–9

Chapter 8
Progress of the Methanol Society

Abstract This chapter presents the world's national actions and technological trends toward a methanol society. While methanol has considerable potential as a fuel, hydrogen carrier, and chemical feedstock, its widespread use remains limited due to economic and policy barriers. A key enabler for a methanol society is the availability of low-carbon methanol synthesized from renewable hydrogen and captured CO_2. However, both hydrogen production—particularly via water electrolysis—and CO_2 capture technologies are currently expensive and energy-intensive. This chapter reviews recent technological advances in these areas, including cost reductions in electrolysis systems and innovations in direct air capture and industrial CO_2 recovery. It also examines emerging pilot projects that integrate renewable energy with methanol synthesis. Despite current limitations, the concept of localized methanol production hubs remains attractive for enhancing energy self-sufficiency and enabling carbon recycling. This chapter emphasizes the need for supportive policy frameworks, targeted investment, and international collaboration. By aligning technological innovation with systemic planning, a methanol-centered approach could offer a realistic and scalable solution to the climate crisis.

Keywords CO_2 capture · Green hydrogen · Electrolysis · Technology development · Carbon neutrality · Infrastructure

8.1 Situation in Major Countries

The countries with the highest CO_2 emissions—*China, the United States*, and *India*—account for more than 50% of global emissions. It is desirable for methanol to become more widely used in these countries. Therefore, we first take a look at the current state of methanol distribution in these three countries, focusing on whether methanol can be quickly introduced as a fuel.

J. Nakamura, M. Ishida, *Blueprint for a Methanol Society*,
https://doi.org/10.1007/978-981-95-3465-4_8

China

China produces approximately 80 million tons of methanol annually, accounting for about 65–70% of global output. China has identified the *methanol economy* as a key element in achieving its carbon-neutral goals. In particular, there is a strong focus on synthesizing methanol using renewable energy sources, which is expected to contribute significantly to the transition to carbon neutrality [1]. As noted earlier, China is actively advancing the practical implementation of methanol as a *chemical feedstock* (notably through the *MTO process*) and as a *transportation fuel* (methanol-fueled vehicles). While methanol is currently produced primarily from coal and natural gas, China is also working to build and diversify its methanol supply chain—an approach that could serve as a reference for other countries. Although the adoption of methanol-fueled vehicles is currently lagging behind the rise of electric vehicles (EVs), the development of methanol refueling infrastructure is underway. In the future, if hydrogen production from renewable energy sources can be fully realized, it will likely make a significant contribution to greenhouse gas reduction [2]. Overall, China is integrating methanol into its broader energy strategy, using it as a tool to achieve environmental targets. If successful, this initiative could serve as a global model for the development of a methanol society.

United States

As of 2025, the United States remains a significant global producer of methanol, leveraging its abundant and low-cost shale gas resources. With an annual domestic production capacity of approximately 10–12 million tons, the United States ranks among the top five methanol-producing countries. Most methanol in the United States is produced using conventional steam reforming of natural gas, and a substantial portion is exported, particularly to Asia, to meet growing demand for methanol-to-olefins (MTO) and fuel blending applications.

Although the adoption of methanol as a direct transportation fuel is still limited within the United States, the country is not necessarily behind in methanol-related developments. Several blue and green methanol projects are underway, including carbon capture-integrated production and renewable hydrogen-based synthesis, supported by federal incentives such as the Inflation Reduction Act (IRA) and Section 45 V hydrogen production tax credits.

The United States does face infrastructure limitations regarding nationwide methanol distribution and refueling networks for marine or automotive use, which may hinder broader domestic fuel adoption. However, methanol is widely used as a chemical feedstock, particularly in formaldehyde, MTBE, and acetic acid production.

Given the scale and integration of its petrochemical and energy sectors, the United States is expected to remain a key player in shaping the future global methanol economy, particularly as low-carbon methanol gains prominence in decarbonization strategies.

The United States is thus expected to be able to rapidly adapt to the future methanol society.

India

As of 2025, India's domestic methanol production is approximately one million tons per year, while an additional 2–three million tons are imported annually, primarily from the Middle East and Southeast Asia. Although India's methanol production infrastructure is still developing, the country has launched an ambitious Methanol Economy program aimed at enhancing energy security, reducing greenhouse gas emissions, and fostering economic growth.

The Indian government, through institutions such as NITI Aayog, is actively promoting methanol as an alternative to imported fossil fuels. Proposed applications include its use in transportation sectors (road, rail, and marine), power generation (diesel generators and industrial boilers), and even domestic cooking. India is also exploring indigenous feedstocks such as high-ash coal, biomass, municipal solid waste, and CO_2 for methanol production to reduce reliance on imports and support circular economy principles.

Methanol and dimethyl ether (DME) are being considered for blending with gasoline and LPG, respectively, to lower fuel costs and reduce local air pollutants. While large-scale commercial deployment is still in its early stages, several pilot projects and R&D initiatives are underway, and policy support continues to strengthen.

With growing demand for clean and affordable energy, methanol is positioned to play an increasingly important role in India's diversified energy portfolio.

Optimistic Expectations

In summary, methanol is in circulation in these three major countries, and it is realistic to think that a large-scale methanol society will be built in the future. Although not covered in detail here, it is important to note that *European countries*—particularly *Denmark, Germany, the Netherlands, and Iceland*—are leading the global push to integrate methanol as a carbon-neutral energy and feedstock solution. Currently, the countries with the largest methanol production volumes are Saudi Arabia, Trinidad and Tobago, Brunei, and Russia.

8.2 Hydrogen Production

The largest challenge in synthesizing methanol from renewable energy sources is the production of hydrogen. As noted earlier, hydrogen production accounts for approximately *70% of the total cost* of methanol synthesis. Here, we will provide an

Table 8.1 Three types of hydrogen produced using different methods

Type of hydrogen	Production methods
Green hydrogen	Produced using renewable energy sources such as solar, wind, or hydropower through water electrolysis. No CO_2 emissions
Blue hydrogen	Produced from natural gas using steam methane reforming (SMR) with carbon capture and storage (CCS) to reduce CO_2 emissions
Gray hydrogen	Produced from natural gas or other fossil fuels using steam methane reforming (SMR) without carbon capture, leading to significant CO_2 emissions

overview of the current state of hydrogen production, with a focus on challenges in realizing a methanol-based society.

In recent years, hydrogen has been categorized according to its production method into green hydrogen, gray hydrogen, and blue hydrogen (see Table 8.1).

- *Green hydrogen* refers to hydrogen produced by *water electrolysis* using electricity generated from renewable energy sources such as solar or wind power.
- *Blue hydrogen* is similar to gray hydrogen, except that the CO_2 by-product is *captured and stored underground* via carbon capture and storage (CCS) technologies.
- *Gray hydrogen* is produced through the reforming of fossil fuels, such as natural gas and coal, with the *CO_2 generated during production being emitted* directly into the atmosphere.

Although other color-based classifications for hydrogen exist, these terms have been gradually adopted and popularized by industry stakeholders, policymakers, researchers, and the media. However, there is no formal consensus on their precise definitions.

For a methanol society that aims to leverage renewable energy, the production of green hydrogen is the most realistic and sustainable pathway. Hydrogen production from biomass also represents a promising technology from a carbon-neutral perspective. While biomass-based hydrogen is unlikely to contribute to large-scale hydrogen supply in the immediate future, it is expected to play an important supporting role in the methanol society of tomorrow.

Challenges and Current State of Hydrogen Production

Let us consider why hydrogen production costs remain high, particularly in the context of electrolysis. There are three primary electrolysis methods (as summarized in Table 8.2): alkaline electrolysis, proton exchange membrane (PEM) electrolysis, and solid oxide electrolysis (SOEC) [3–6].

- *Alkaline electrolysis* uses an aqueous alkaline solution as the electrolyte and is the most mature and widely used technology. It offers system efficiencies of

Table 8.2 Hydrogen production by electrolysis using AES, PEC, and SOEC [3–5]

Type	Overview	Energy efficiency	Advantages	Disadvantages
Alkaline electrolyzer cell (AEC)	Alkaline electrolyte (KOH or NaOH)	70–80%	Well-established, long operational history, relatively low-cost, large-scale hydrogen production	Lower efficiency due to low current density. Long startup time and less adaptable to rapid load changes
Proton exchange membrane (PEM)	Solid polymer membrane. Electrolyte, proton carrier (H^+).	60–80%	High current density, compact device design, quick startup and shutdown, efficient hydrogen production	High initial cost due to the need for expensive catalysts like iridium and platinum. Membrane has a finite lifespan
Solid oxide electrolyzer cell (SOEC)	High temperatures (700–1000 °C), oxide ion carrier (O^{2-})	80–90% (theoretical)	Very high efficiency achievable with potential use of excess heat. High-purity hydrogen production	High-temperature operation, material durability, very high initial cost, risk of material degradation at high-temperature operation

approximately 70–80% (based on higher heating value, HHV), but its response to intermittent renewable electricity is limited.

- *Proton exchange membrane (PEM) electrolysis* uses a solid polymer membrane and enables compact, flexible systems with higher current density. It achieves system efficiencies of about 67–82% today, with projections reaching 82–86% by 2030.
- *Solid oxide electrolysis (SOEC)* operates at high temperatures (typically 600–900 °C) and has the highest theoretical efficiency—over 90%—when integrated with waste heat. However, SOEC remains in the early commercial or advanced research stage, with challenges related to material durability and cost.

The high cost of electrolysis stems primarily from capital expenditures. Installing electrolyzers and renewable energy infrastructure involves substantial upfront investment, including electrolyzer stacks, power electronics, thermal management, and gas handling systems.

Key cost drivers include:

- *Electrocatalysts*, such as iridium (for oxygen evolution) and platinum (for hydrogen evolution), which are expensive and scarce. As of 2024, iridium costs approximately US$130 per gram and platinum around US$32 per gram.

- *Membranes*, especially Nafion® used in PEM systems, cost approximately US$500 per square meter due to their excellent chemical stability and high proton conductivity.
- *Cell stacks*, composed of multiple electrochemical cells, account for roughly 30% of the system cost, in addition to supporting components like cooling units and gas purification systems.
- *Operating costs*, particularly electricity, remain significant. Even when powered by renewable energy, variability and supply stability pose challenges. Maintenance costs, especially for replacing expensive membranes and catalysts, also contribute.

Thus, hydrogen production remains a major bottleneck not only for a methanol society but for the broader hydrogen economy. As of 2025, water electrolysis remains the only viable method to produce green hydrogen at scale. Although direct electrochemical conversion of CO_2 to methanol or other fuels (e.g., formic acid and ethanol) is under active research, these technologies are not yet ready for industrial deployment.

Hydrogen Production from Biomass

Hydrogen production from *biomass* is increasingly viewed as a promising approach [7–9]. Biomass resources include:

i. *Agricultural and forestry residues* (e.g., rice straw, corn stalks, wood chips, and bark)
ii. *Energy crops* (e.g., miscanthus, switchgrass, and sugarcane)
iii. *Lignocellulosic biomass* (e.g., wood, rice straw, and corn leaves).

Biomass offers the advantage of carbon-neutral hydrogen production, as it absorbs CO_2 during growth. However, energy inputs for high-temperature conversion processes can lead to additional CO_2 emissions if fossil fuels are used.

Biomass-derived hydrogen production processes can be broadly divided into three categories:

- *Thermochemical processes*: including *gasification* (reacting biomass with oxygen or steam at 700–1000 °C to produce synthesis gas) and *pyrolysis* (thermal decomposition at 300–700 °C). These methods are efficient but energy-intensive.
- *Biological processes*: such as *fermentation* (microbial conversion of sugars to hydrogen) and *photosynthetic hydrogen production* (using algae or photosynthetic bacteria). While these processes have lower environmental impact and operate at lower temperatures, they suffer from slow production rates.
- *Chemical processes*: reforming *liquid biofuels* (e.g., methanol and ethanol) to produce hydrogen. While convenient for storage and transport, using methanol or ethanol as fuels directly is typically more practical.

Regional Perspectives

The potential for biomass-based hydrogen production varies by region:

- *China*: The world's largest agricultural producer, generating vast amounts of agricultural waste. Robust biomass utilization infrastructure, supported by the government's "Carbon Neutral 2050" policy, strengthens the feasibility of biomass-based hydrogen.
- *India*: Abundant agricultural residues and growing renewable energy adoption create opportunities for biomass-based hydrogen, though rural infrastructure development is essential.
- *Southeast Asia*: Countries like Indonesia, Malaysia, and Thailand have abundant palm oil waste and agricultural residues. However, environmental impacts, such as deforestation, must be carefully managed.
- *United States*: With major agricultural production and vast forestry resources, there is great potential for utilizing agricultural and forestry by-products as feedstocks for hydrogen production. Urban food waste can also be leveraged through fermentation processes.

Hydrogen production from biomass is generally less expensive than electrolysis. Regions in Asia, especially Southeast Asia and China, with abundant biomass resources and supportive policies, are well-positioned to realize large-scale biomass-based hydrogen production. Technological innovation and infrastructure development in these regions could establish a sustainable hydrogen economy.

In summary, the most realistic hydrogen production pathways over the next 30 years are water electrolysis and biomass utilization. Both approaches will be critical for enabling a methanol-based society. In the early phase, gray hydrogen from fossil fuels may be necessary to build up industrial infrastructure. Among the alternatives, biomass-based hydrogen shows great promise in Southeast Asia and China due to its resource base and policy environment. Ultimately, building a methanol value chain will require coordinated global strategies, including regulatory incentives and innovation to drive down costs and support widespread adoption.

8.3 Carbon Dioxide Recovery Technologies

In the production of green methanol, approximately 70% of the cost is attributed to the generation of green hydrogen, while around 10% is attributable to the cost of procuring CO_2 feedstock. Currently, the majority of recovered CO_2 originates from fossil fuel-derived sources.

CO_2 Capture Technologies

As shown in Table 8.3, the principal technologies for CO_2 capture include chemical absorption, physical adsorption, membrane separation, and chemical looping [10–15]:

- *Chemical Absorption*: This mature and widely used method employs aqueous amine-based solvents to absorb CO_2 from flue gases. The CO_2-rich solvent is then thermally regenerated to release concentrated CO_2. While this approach offers high capture efficiency (typically >90%), it requires significant energy input and presents challenges such as solvent degradation, corrosion, and high operating costs.
- *Physical Adsorption*: In this method, solid sorbents capture CO_2 through physisorption. Desorption is achieved through pressure or temperature swings (PSA/TSA). Advantages include low-temperature operation and high selectivity depending on the sorbent material. However, long-term stability and sorbent regeneration costs remain critical issues.
- *Membrane Separation*: This technology uses selective membranes—either polymeric or inorganic—to separate CO_2 from gas mixtures. Advantages include continuous operation and a compact system design. Challenges involve membrane fouling, limited selectivity, and membrane replacement costs. Membrane fouling refers to the accumulation of materials on membrane surfaces or within pores, reducing separation performance over time.
- *Chemical Looping*: An emerging CO_2 capture technology based on cyclic redox reactions of metal oxides. In this process, fuel is combusted with a metal oxide instead of air, producing a CO_2-rich exhaust stream without nitrogen dilution.

Table 8.3 Various CO_2 capture methods

Technology	Method	Process	Advantages	Challenges
Absorption	Absorbing CO_2 into a chemical absorbent solution	Use amine solution. Separate CO_2 by heating	High-efficiency CO_2 capture	Corrosiveness of amines and high regeneration cost
Adsorption	Adsorbing CO_2 using solid adsorbents. Separating by heating or depressurizing	Pass CO_2 through an adsorption tower, then separate CO_2	Operable at low temperatures, high selectivity	Deterioration, cost of adsorbents, reduced adsorption efficiency
Membrane separation	Selectively separating CO_2 through a membrane	Use polymer or inorganic membranes to pass CO_2	Continuous operation, compact equipment	Durability and permeability of membranes, cost of membranes
Chemical looping	Fuel combustion using metal oxides as oxygen carriers	Producing a concentrated CO_2 stream suitable for capture	High temperatures, high-efficiency separation	Lifespan, cost of materials, complexity of operation

The reduced metal oxide is then regenerated by oxidation with air. While this method offers high-purity CO_2 and integrated energy recovery, it remains at an early stage of commercialization due to material stability and reactor design complexity.

Capture Efficiency and Economic Considerations

Modern CO_2 capture technologies can achieve efficiencies exceeding 90%. However, economic viability depends on factors such as capital expenditures, operating costs, energy consumption, and scale.

- *Chemical absorption* remains the benchmark for large-scale capture, but its regeneration step consumes significant energy. Capital costs can reach several hundred million dollars, and operating costs typically range from *US$40 to US$100 per ton of CO_2 captured*, depending on solvent type and process integration.
- *Physical adsorption* systems generally have lower capital costs, but operational costs vary widely from *US$50 to US$150 per ton*, reflecting sorbent degradation and regeneration energy.
- *Membrane separation* offers moderate costs and modular scalability. Current estimates place operating costs at *US$23 to US$50 per ton*, though system efficiency and membrane durability are crucial.
- *Chemical looping*, while promising for integrated CO_2 capture and heat generation, remains in development. Projected operating costs range between *US$30 and US$70 per ton*, depending on the oxygen carrier material and reactor scale.

Economies of scale are vital for reducing per-unit costs, and ongoing technological innovation is expected to enhance the competitiveness of these methods.

Direct Air Capture (DAC)

Direct Air Capture (DAC) is a rapidly developing technology that removes CO_2 directly from ambient air. Two main approaches are under investigation:

- *Liquid solvent systems*: Air is passed over alkaline solutions (e.g., potassium hydroxide), forming carbonate compounds that are subsequently processed to release pure CO_2 and regenerate the solvent.
- *Solid sorbent systems*: Air passes over solid materials (e.g., amine-functionalized adsorbents) that bind CO_2, which is later released via heating or pressure swing.

Leading companies include *Climeworks* (Switzerland), *Carbon Engineering* (Canada), and *Global Thermostat* (USA). Currently, DAC is expensive, with estimated costs ranging from *US$500 to US$1000 per ton* for small-scale operations.

However, recent studies suggest that DAC at scale (≥1 Mt./year) could reduce costs to between *US$94 and US$232 per ton* by 2030, assuming technological improvements and cost reductions in materials and energy inputs.

Carbon Capture and Storage (CCS) Initiatives

Carbon capture and storage (CCS) involves the permanent geological storage of CO_2. As shown in Table 8.4, several notable projects are underway around the world:

- *United States*: Approximately *12 million tons* of CO_2 are captured annually. Significant support is provided by the Department of Energy and the private sector.

Table 8.4 CO_2 capture project

Country	Key projects	Government support	Advantages
United States	Petra Nova project: One of the world's largest post-combustion CCS projects, ~1.4 million tons/year of CO_2; Illinois industrial CCS project: ~one million tons/year of CO_2 from an ethanol production facility	Federal funding and tax credits; DOE'S strong push for CCS research and development	Large-scale demonstration projects; strong federal support for commercialization
Norway	Sleipner project: World's first commercial-scale CCS project, capturing ~one million tons/year of CO_2; northern lights project: Large-scale CO_2 storage from across Europe in the North Sea	Significant investments in CCS technology; strong regulatory support and incentives	Long-standing stable operation of CCS projects; leadership in subsea storage technology
China	Ordos CCS project: Shaanxi Province; Yanchang CCS project: Demonstration project at a petrochemical plant, capturing ~400,000 tons of CO_2 annually	National policy support and funding for CCS technology development and demonstration	Large-scale projects with significant CO_2 capture capacity; strong government backing
Canada	Quest CCS project: Over one million tons/year of CO_2 from oil sands upgrading; boundary dam project: World's first full-scale CCS project at a coal-fired power plant	Provincial and federal funding; strong regulatory framework and support for CCS projects	Successful large-scale projects; strong regulatory and policy framework
England	Teesside and Humber projects: CO_2 capture and storage projects in industrial areas; acorn CCS project: CCS project using existing oil and gas infrastructure in the North Sea	Policy support for climate change mitigation; funding for CCS commercialization	Utilization of existing industrial clusters and infrastructure; strong policy support

- *Norway*: The *Sleipner* project has been in operation since 1996, injecting CO_2 into a saline aquifer in the North Sea.
- *China*: Captures approximately *three million tons* of CO_2 per year, primarily from industrial facilities.
- *Canada*: The *Boundary Dam* project demonstrates CCS in coal-fired power generation and is linked to the *Alberta Carbon Trunk Line*.
- *United Kingdom*: Projects such as *Zero Carbon Humber* are developing CCS infrastructure using depleted North Sea reservoirs.
- *Australia*: The *Gorgon* project stores about *four million tons* of CO_2 annually from gas processing operations.

Despite these efforts, global CO_2 capture accounts for *less than 0.1%* of total annual emissions as of 2023, underscoring the need for further deployment and policy support.

Policy Mechanisms: The U.S. 45Q Tax Credit

In the United States, the *Section 45Q tax credit* supports CCS and CCU projects by offering:

- *US$50 per ton* of CO_2 permanently stored in geological formations.
- *US$35 per ton* of CO_2 used for commercial purposes, such as enhanced oil recovery (EOR).

Eligible projects must meet minimum capture thresholds and comply with EPA regulations. The credit applies for up to *12 years* from the project start date and has significantly improved the economics of CCS.

International Incentives and Support

Globally, multiple policy frameworks support CO_2 capture, utilization, and storage:

- *European Union*: The *EU Emissions Trading System (ETS)* offers carbon pricing, and the *Innovation Fund* supports demonstration-scale CCUS projects.
- *Norway*: Government funding facilitates projects such as *Northern Lights*, which aims to build transnational CO_2 transport and storage infrastructure.
- *Canada*: Federal and provincial tax incentives, particularly in Alberta, promote CCS deployment.
- *United Kingdom*: The *Carbon Price Support mechanism* and the *CCUS Infrastructure Fund* provide financial support for industrial CCS.
- *Australia*: The *Emissions Reduction Fund* enables CCS projects to earn credits and participate in carbon markets.

These incentives are essential to overcoming the technical and economic barriers to large-scale CO_2 capture and enabling the transition toward a low-carbon economy.

References

1. Huy Q (2022) Could Green Methanol Be What China Needs to Reach Net Zero? INSEAD Knowledge. https://knowledge.insead.edu/strategy/could-green-methanol-be-what-china-needs-reach-net-zero
2. Irena and Methanol Institute (2021) Innovation outlook: renewable methanol. International Renewable Energy Agency, Abu Dhabi. https://www.irena.org/Publications/2021/Jan/Innovation-Outlook-Renewable-Methanol
3. Ursu A, Gandı LM, Sanchis P (2012) Hydrogen production from water electrolysis: current status and future trends. Proc IEEE 100:410–426
4. Kumarab SS, Lim H (2023) Recent advances in hydrogen production through proton exchange membrane water electrolysis—a review. Sustain Energy Fuels 7:3560–3583
5. Sapountzi FM, Gracia JM, Weststrate CJ, Fredresksson HOM, Niemantsverdiet JW (2017) Electrocatalysts for the generation of hydrogen, oxygen and synthesis gas. Prog Energy Combust Sci 58:1–35
6. IRENA (2020) Green hydrogen cost reduction: scaling up Electrolysers to meet the 1.5°C climate goal. International Renewable Energy Agency, Abu Dhabi
7. Ni M, Leung DYC, Leung MKH, Sumathy K (2006) An overview of hydrogen production from biomass. Fuel Process Technol 87(2006):461–472
8. Parthasarathy P, Narayanan KS (2014) Hydrogen production from steam gasification of biomass: Influence of process parameters on hydrogen yield–a review. Renewable Energy 66:570–579
9. Gonzalez-Garay A, Frei MS, Al-Qahtani A, Mondelli C, Guillen-Gosalbez G, Perez-Ramırez J (2019) Plant-to-planet analysis of CO_2-based methanol processes. Energy Environ Sci 12:3425–3436
10. Zhang XG, Abdul Raman AA, Jewaratnam J, Buthiyappan A (2024) Sustainable carbon dioxide capture, storage, and utilization: review of current status and future direction. Int J Environ Sci Technol 22:6125. https://doi.org/10.1007/s13762-024-05908-x
11. Charalambous C, Moubarak E, Schilling J, Fernandez ES, Wang JY, Herraiz L, Mcilwaine F, Peh SB, Garvin M, Jablonka KM, Moosavi SM, Herck JV, Ozturk AY, Pourghaderi A, Song AY, Mouchaham G, Serre C, Reimer JA, Bardow A, Smit B, Garcia S (2024) A holistic platform for accelerating sorbent based carbon capture. Nature 632:89–95
12. Powell CE, Qiao GG (2006) Polymeric CO_2 /N_2 gas separation membranes for the capture of carbon dioxide from power plant flue gases. J Membr Sci 279:1–49
13. Li PF, Xu Y, Chen QB (2024) Global trends and research characteristics on CO_2 capture and conversion from 2000 to 2023: a bibliometric review. Sustain Energy Technol Assess 67–103836:1–12
14. Rochelle GT (2009) Amine scrubbing for CO_2 capture. Science 325:1652–1654
15. Sanz-Pérez ES, Murdock CR, Didas SA, Jones CW (2016) Direct capture of CO_2 from ambient air. Chem Rev poiu116:11840–11876

Part III
What Should We Do? Business, Research and Development, Policy, Philosophy, and Education

Chapter 9
Road Map and Business Development of the Methanol Society

Abstract This chapter outlines a comprehensive road map for transitioning to a methanol-based society, presenting milestones in technology development, infrastructure deployment, and policy alignment. The road map envisions a phased approach: in the short term, methanol adoption will rely on fossil-based CO_2; in the mid-term, industrial CO_2 capture and renewable hydrogen will scale up; and in the long term, a fully circular carbon economy will emerge using direct air capture and green methanol synthesis. Central to this transition is the establishment of robust methanol supply chains—including CO_2 sources, hydrogen production, methanol synthesis, storage, transport, and end-use infrastructure. The development of viable methanol businesses is also emphasized as a driver for investment, innovation, and international trade. Creating a value chain that connects CO_2 emitters, renewable energy providers, and chemical or fuel consumers is critical. This chapter introduces the concept of "methanol hubs" to concentrate resources and reduce logistical burdens. Success will depend on coordinated policy support, public-private partnerships, and measurable implementation targets. The road map highlights the need for system-level integration to ensure that the methanol economy becomes both technically viable and economically sustainable.

Keywords Road map · Supply chain · Methanol business · Infrastructure development · Methanol hubs · Energy transition

9.1 Road Map for Building a Methanol Society

Urgent Priority: CO_2 Recovery

The key actions that should be initiated immediately to establish a methanol-based society are summarized in Table 9.1. The most urgent priority is the recovery of CO_2. Since methanol is synthesized from CO_2 and H_2, securing a stable supply of CO_2 is essential. Abruptly halting the use of fossil resources is not realistic;

J. Nakamura, M. Ishida, *Blueprint for a Methanol Society*,
https://doi.org/10.1007/978-981-95-3465-4_9

Table 9.1 Urgent priorities for building a methanol society

Priority	Description	Urgency
CO_2 capture from high-concentration emission sources	Capture of CO_2 from boilers in thermal power plants, steel mills, cement plants, chemical plants, etc.	Very high
Methanol synthesis from CO_2	Urgent construction of plants to synthesize methanol from CO_2	High
Utilization of methanol as fuel	Use of methanol as a fuel for internal combustion engines in thermal power generation or fuel cells	High
Large-scale hydrogen production	Infrastructure development necessary for large-scale hydrogen production	High

therefore, the first step should be to mitigate global warming by recovering as much CO_2 as possible.

Construction of Methanol Synthesis Plants and Acceleration of Methanol-Compatible Technologies

The next priority is the construction of small-and medium-scale methanol synthesis plants. Although engines and fuel cells that use methanol as fuel already exist, their production must be accelerated. Furthermore, efforts to produce hydrogen at relatively low cost must begin immediately. As previously discussed, hydrogen production currently accounts for approximately 70% of the cost of methanol production. This cost reflects a significant portion of the energy input because water electrolysis for hydrogen production requires substantial electrical energy. While photocatalytic methods for hydrogen production exist, they are not practical for large-scale production. Realistically, electrolysis is currently the only viable method for producing large quantities of hydrogen.

Hydrogen Demand and the Need for Immediate Action

To put this in perspective, the amount of hydrogen equivalent to the energy of fossil fuels associated with annual CO_2 emissions of 35 billion tons is estimated to be 4.5 billion tons of hydrogen (see Sect. 4.3). While current targets from the United States (10 Mt./year) and EU (20 Mt./year) are significant, they still fall far short of the several hundred million tons per year estimated for a full-scale methanol society. Therefore, rather than waiting for the emergence of a new hydrogen production method, we need to respond using currently available technologies.

The Need for Stepwise Development

While rapidly implementing these immediate priorities, it is also necessary to plan for the stepwise development of industries, businesses, science and technology, and policy measures, in order to guide the methanol society toward maturity.

Road Map for the Methanol Society: Three Phases

In this chapter, we propose a road map to achieve this transition. We divide the development of a methanol-based society into three phases and present a road map accordingly.

- **Phase 1 (2025–2030)** represents the birth period of the methanol society, during which methanol fuel is introduced alongside continued use of fossil resources.
- **Phase 2 (2031–2040)** is the growth period of the methanol society, characterized by a global transition to methanol-based systems.
- **Phase 3 (2041–2050)** represents the maturity period of the methanol society, during which we will have developed a comprehensive understanding of how humanity interacts with energy and materials.

While 2025–2030 is proposed as the birth phase, the actual pace of adoption will depend heavily on policy support, technology deployment, and public acceptance.

9.2 Phase 1 (2025–2030): The Birth of the Methanol Society

Phase 1 (2025–2030) marks the birth of the methanol society. This period is critical for establishing foundational infrastructure and systems, requiring decisive policy actions, promotion of CO_2 recovery, creation of methanol markets, development of infrastructure, and enhanced public understanding. While carbon pricing has begun to show effects in the EU, progress remains limited in major emitting countries like China (31.7%), the United States (12.7%), India (7.9%), and Russia (5.1%). Achieving a broad international consensus and ensuring public support are essential. Table 9.2 summarizes the key activities and goals for Phase 1.

Industrial Focus and the Global Methanol Value Chain

In the industrial sector, the priority must be on expanding methanol production while simultaneously promoting CO_2 recovery, laying the groundwork for a next-generation methanol market. Building a global methanol value chain is also

Table 9.2 Phase 1 of the road map for a methanol society

Phase 1: Initial Deployment and Foundation Building (2025–2030)
Key activities and goals
1. Foundation and policy initiatives • Set clear policies and secure funding • Launch methanol markets and infrastructure projects
2. Industrial and hydrogen expansion • Expand hydrogen and methanol production • Promote biomass-derived hydrogen technologies
3. High-concentration CO_2 recovery • Recover CO_2 from industrial sources • Use it for methanol synthesis
4. Infrastructure and rising costs • Install methanol stations and pipelines • Secure policies to offset rising costs
5. Outreach, education, and engagement • Run public education campaigns • Engage citizens and professionals actively

imperative. Over one-third of methanol consumption already goes to fuel applications, and methanol-compatible engines and infrastructure are in place, giving methanol an advantage over other carriers like hydrogen, ammonia, and methylcyclohexane (MCH).

Expansion of Hydrogen Production and Use of Biomass

It is essential to promote diversified hydrogen production methods in parallel with the introduction of existing CO_2 capture technologies. Among them, hydrogen from biomass is particularly promising, and gasification of waste biomass and agricultural residues is a top priority. The use of blue hydrogen, typically produced from natural gas with carbon capture and storage (CCS), needs to be promoted. Meanwhile, fossil-derived gray hydrogen will remain necessary at this early stage to support the expansion of the large-scale methanol market.

Scaling Up Water Electrolysis

Scaling up hydrogen production via electrolysis is another critical priority. Despite the high costs of precious metal catalysts, water electrolysis remains a key method. Improvements in electrolyzer efficiency, development of high-performance, non-precious metal catalysts (like nitrogen-doped carbon), and advances in SOEC and PEM technologies will be crucial for this transition.

High-Concentration CO_2 Recovery and Methanol Synthesis

High-concentration CO_2 recovery from thermal power plants, steel mills, cement plants, and chemical plants is an immediate priority. Instead of underground storage, synthesizing methanol from recovered CO_2 offers a practical alternative. CO_2 recovery technologies are already established (see Sect. 8.3).

Infrastructure Development and Deployment of Methanol Stations

From an infrastructure perspective, the strategic deployment of methanol stations is essential. Since methanol can be reformed to produce hydrogen, reformers should be installed upstream of hydrogen fuel cells (see Sect. 10.4). Construction of pipelines for hydrogen, methanol, and CO_2, with rigorous safety standards, will connect production and consumption centers.

Rising Energy Costs and the Need for Policy Support

A crucial point to consider is the **rising energy costs**. Fossil fuels have been historically cheap, but methanol synthesized from hydrogen and recovered CO_2 will inevitably be more expensive. Robust policy support and educational initiatives will be essential to address these challenges, as carbon neutrality cannot be achieved through market forces alone.

Policy Measures and Market Incentives

From a policy standpoint, government subsidies and tax incentives are vital to lower the initial investment burden for hydrogen production, infrastructure, and fuel cell vehicle adoption. Carbon pricing, including carbon taxes and emissions trading systems, will also be essential for creating economic incentives to accelerate clean hydrogen production and reduce emissions.

The Role of Public Understanding and Civic Engagement

Policy discussions must be transparent and informed by scientists and engineers. Dialogue with experts from other fields and engagement with ordinary citizens are key to addressing climate change. Public awareness is the most effective driver of climate action, and social media can be leveraged to reach younger generations.

Outreach, Education, and Demonstration Projects

Finally, promotion and outreach efforts play a critical role. Public campaigns, educational programs, and regional demonstration projects should communicate the benefits of methanol and hydrogen, build consumer trust, and inspire adoption in other regions.

Summary of Phase 1 Objectives

In summary, Phase 1 (2025–2030) establishes the foundation of the methanol society by prioritizing CO_2 recovery, hydrogen production, and the initial expansion of methanol use across industries. Key strategies include: Scaling up hydrogen production, including biomass-based and electrolysis-derived hydrogen, establishing CO_2 recovery infrastructure and methanol supply chains, developing methanol-compatible infrastructure and reformers, building robust policy support and economic incentives, and promoting public awareness and engagement. Success in Phase 1 is essential to prepare for the growth and maturation of the methanol society in subsequent phases, driving the transition toward a sustainable and low-carbon future.

9.3 Phase 2 (2031–2040): Growth and Expansion of the Methanol Society

Phase 2 (2031–2040) marks the growth period of the methanol society. During this phase, the focus will be on the maturation of technology, strengthening of supply chains, and widespread market introduction. The aim is to expand the methanol economy and establish it as a core element of a sustainable energy system. Table 9.3 summarizes the key activities and goals for phase 2.

Table 9.3 Phase 2 of the road map for a methanol society

Phase 2: Growth and Expansion (2031–2040)
Key activities and goals
1. Renewable energy expansion and integration • Build large-scale solar and wind power plants • Enable power-hydrogen hybrid systems for methanol synthesis
2. Industrial applications and CCU technologies • Promote methanol in steel and chemical industries • Integrate CCU and methanol-based hydrogen supply
3. Global methanol market development • Foster international methanol trade and cooperation • Standardize methanol-related technologies globally
4.Technology maturation and cost reduction • Construct large-scale hydrogen production • Automate and optimize processes for efficiency
5. Outreach and social acceptance • Launch public campaigns and education programs • Showcase demonstration projects in industrial clusters

Expansion of Renewable Energy Utilization

A critical first step is the expansion of renewable energy use. Large-scale solar and wind power plants will be constructed, and the electricity generated from these sources will be utilized for hydrogen production, which serves as the essential feedstock for methanol synthesis. To address the intermittency of renewable energy, power-hydrogen hybrid systems will be deployed, integrating energy storage systems to ensure a stable hydrogen supply. Additionally, small-and medium-sized methanol production facilities will be installed in proximity to power plants to streamline hydrogen transportation.

Expansion of Industrial Applications

The growth of industrial applications is another key element in Phase 2. In the steel industry, methanol-based direct reduction of iron is under active investigation, with pilot-scale demonstrations in development. In the chemical industry, methanol will substitute for fossil fuels, while CO_2 capture and utilization (CCU) technologies will recover CO_2 emissions from industrial processes and power plants. By reacting captured CO_2 with hydrogen, methanol produced from renewable energy sources will become more prevalent. Although hydrogen-based steelmaking is being explored globally, methanol reforming offers a simpler and more flexible method to supply hydrogen.

Establishing a Global Methanol Market

Building a global market for methanol is indispensable. The international circulation of methanol, CO_2, and hydrogen must be promoted, with advanced economies playing a leading role. In Europe, methanol is already being adopted as a carbon-neutral fuel—for example, by shipping giant *Maersk*, which has ordered over 20 vessels powered by green methanol. Port hubs such as *the Port of Rotterdam* and *Aarhus Port* are developing bunkering infrastructure to support this transition. In Asia, *China's green methanol export projects*, particularly in Inner Mongolia and Xinjiang, are rapidly scaling up production capacities with plans for overseas shipments.

The knowledge and technologies developed in these regions should be disseminated worldwide. International cooperation and the standardization of methanol-related technologies and products—through organizations such as *the Methanol Institute* and initiatives like *the International Maritime Organization's (IMO) Alternative Fuels Agenda*—are essential for ensuring compatibility and interoperability. Countries should jointly pursue hydrogen-related research and development (R&D) projects, such as *Horizon Europe's Clean Hydrogen Partnership* and *Japan's Green Innovation Fund*, to accelerate progress and share technological advancements. Furthermore, international transportation infrastructure for methanol and associated gases should be developed, supported by bilateral trade agreements and multilateral frameworks to foster the growth of a robust global methanol market. International transport of CO_2 may face regulatory challenges and requires bilateral agreements.

Technology Maturation and Cost Reduction

Achieving the maturation of technology and reducing costs are central goals of this phase. This includes constructing large-scale hydrogen production facilities to leverage economies of scale and reduce costs. Hydrogen production processes should be automated and optimized to improve efficiency, while long-term durability testing of production equipment and infrastructure will be conducted to enhance reliability. Implementing preventive maintenance and regular maintenance programs will help to reduce operating costs and improve system performance.

Promoting Dissemination and Social Acceptance

Equally important is promoting the dissemination of methanol-based solutions and improving social acceptance. Public information campaigns will be launched to highlight the advantages and success stories of the methanol society, thereby gaining broader consumer understanding and support. Educational programs in schools and universities will be expanded to spread knowledge about methanol's role in a sustainable energy future. Large-scale demonstration projects in key industrial clusters will showcase the benefits of methanol carriers, contributing to local economic revitalization and environmental improvement in partnership with local communities.

Summary of Phase 2 Objectives

In summary, Phase 2 (2031–2040) aims to solidify the growth and expansion of the methanol society through the maturation of technology, supply chain development, and broad market adoption. Key pillars include: Expansion of renewable energy use, growth of industrial applications, development of a global methanol market, technology maturation and cost reduction, and promotion of dissemination and public engagement. By successfully implementing these strategies, the methanol society will be poised for the maturity and full-scale deployment anticipated in Phase 3.

9.4 Phase 3 (2041–2050): Maturity of the Methanol Society

Phase 3 (2041–2050) marks the maturity phase of the methanol society. This period is crucial for establishing methanol's role across all sectors, requiring the full deployment of technology and infrastructure, widespread utilization, and maturation of policies and markets. Table 9.4 summarizes the key activities and goals for Phase 3.

Full Deployment of Methanol in all Sectors

A primary focus in Phase 3 is the large-scale implementation and dissemination of methanol technologies. In the transportation sector, further deployment of methanol fuel cell vehicles (FCVs) and methanol fuel hybrid vehicles will be pursued. Additionally, methanol fuel will be introduced to heavy-duty transportation modes, including trucks, railways, and ships.

In residential and commercial sectors, home and commercial fuel cell systems powered by methanol will be installed for heating and electricity generation. In the industrial sector, the goal is complete decarbonization of processes by replacing fossil fuels entirely with hydrogen and methanol-derived fuels. Key industrial processes—including steel, chemical, and cement manufacturing—will aim for net-zero CO_2 emissions. Furthermore, combining biomass-derived carbon with hydrogen will expand the production of biofuels and bioplastics.

Table 9.4 Phase 3 of the road map for a methanol society

Phase 3: Maturity and Widespread Adoption (2041–2050)
Key activities and goals
1. Full deployment in all sectors • Expand methanol FCVs and hybrid vehicles • Achieve net-zero CO_2 in key industries
2. Cost reduction and technological optimization • Lower methanol production and use costs • Integrate with smart grid for dynamic supply
3. Sustainable and circular energy system • Promote CCU and negative emissions • Use waste-based hydrogen for circular economy
4. International cooperation and market maturity • Expand global trade and standardization • Collaborate in R&D and capacity-building efforts
5. Social acceptance and outreach • Run continuous campaigns to promote methanol • Incorporate methanol in education and demonstration

Cost Reduction and Technological Optimization

Another critical priority is the reduction of costs and the optimization of technologies. This includes:

i. Operational optimization of large-scale manufacturing facilities to lower methanol production costs
ii. Improvement of efficiencies throughout the production, transportation, storage, and end-use phases of methanol to reduce overall energy and operational costs
iii. Introduction of new hydrogen production technologies, such as high-temperature electrolysis and photocatalytic methods, to further boost efficiency

It is also essential to integrate methanol-based energy systems with smart grid infrastructures to balance energy supply and demand dynamically.

Toward a Sustainable and Circular Energy System

A key objective of Phase 3 is the establishment of a sustainable energy system. This includes achieving complete carbon neutrality across all industries and regions by integrating methanol-based energy systems. Technologies such as carbon dioxide capture and utilization (CCU) and negative emission technologies will be promoted. Additionally, hydrogen production methods using waste materials as feedstock will contribute to a resource-circulating economy. Continued expansion of renewable resource-based hydrogen production will reinforce the sustainability of this system.

International Cooperation and Market Maturity

International cooperation and market maturity are also essential. Expanding international trade of methanol and methanol-derived products will help establish a global supply chain. Standardizing international methanol energy technologies and product specifications will facilitate compatibility and cross-border transactions.

Joint R&D initiatives and collaborative technological innovation will be crucial for global progress. Technology transfer and capacity-building support for developing countries will accelerate the global adoption of methanol energy systems and promote a worldwide methanol society.

Promoting Social Acceptance and Outreach

Equally important is social acceptance and outreach. Continuous public campaigns will highlight the advantages and safety of methanol energy. Incorporating methanol energy topics into educational curricula at schools and universities will cultivate a next generation of experts. Large-scale, nationwide demonstration projects will validate the practicality of methanol energy and provide successful case studies that can encourage adoption in other regions.

Summary of Phase 3 Objectives

In summary, Phase 3 (2041–2050) targets the maturity and widespread dissemination of the methanol society. This requires: Full deployment of technology and infrastructure, widespread use in all sectors, and maturity of policy frameworks and markets. Success during this phase is essential for realizing a sustainable energy system and a circular economy. By fostering cooperation among stakeholders and introducing innovative technologies and efficient infrastructure, the vision of a true methanol society will become a reality.

9.5 Methanol Business: Driving the Next Industrial Revolution

Linking Methanol to the Global Economy

In the birth and growth of the methanol society, a critical challenge lies in integrating methanol into the global economy and making it relevant to the lives of citizens worldwide. The ongoing development of hydrogen production, renewable energy, and the methanol economy can be viewed as a transformation on par with the Industrial Revolution. To rapidly reduce CO_2 emissions, it is essential to launch and expand businesses focused on CO_2 capture and methanol production. Policy support through subsidies and incentives will play a key role in accelerating this transformation. The methanol sector is expected to generate a wide array of business opportunities over the next 10 to 20 years.

Governmental Policies and Regulatory Frameworks

To advance this energy revolution, coordinated efforts are needed across policy, corporate interests, social well-being, and future sustainability. Proactive government policy is essential. Governments should expand investments in technologies for methanol conversion, CO_2 recovery, hydrogen production, methanol synthesis, heat utilization, and the entire methanol value chain. Additionally, they must offer regulatory relief and incentives—such as subsidies and tax breaks—to enable companies to adopt these new technologies. Robust regulations governing the production, transportation, and use of methanol are also necessary, with careful attention to safety and environmental impacts. Stricter CO_2 emissions regulations will accelerate the transition to clean energy, improving the sustainability of the entire industry while reducing environmental footprints.

Corporate Strategies and Cross-Industry Collaboration

For companies, it is crucial to recognize that transitioning to a sustainable energy system offers long-term economic and strategic benefits. While initial investments may be significant, they will yield energy cost savings, regulatory risk reduction, and access to new markets. Above all, this transition enhances corporate reputation, winning public support and attracting top talent. Industries such as energy, chemicals, transportation, and automotive must collaborate and share a unified vision to build an efficient methanol economy. Cross-industry cooperation will form effective value chains, while partnerships and joint technological development will drive innovation. Furthermore, companies should integrate environmental stewardship into their corporate social responsibility (CSR) initiatives, contributing to the development of a sustainable society.

Societal Impact and Workforce Adaptation

As global temperatures continue to rise, societies will increasingly recognize the urgency of the problem, inevitably triggering an energy and industrial revolution. The adoption of methanol-related technologies will create new jobs and industries, spurring regional economic revitalization. However, technological progress will also pose challenges, including the decline of existing jobs. Therefore, governments and companies must implement reskilling and vocational training programs to help workers adapt and thrive in this new industrial landscape.

Vision for a Sustainable Future

From a future perspective, this industrial revolution represents a crucial step toward realizing a sustainable society. Expanding renewable energy use and constructing a methanol economy will reduce dependence on fossil fuels and curb global warming. Furthermore, sustainable resource management and environmental protection will ensure that a healthier planet is passed on to future generations. With ongoing technological innovation, significant advances are also expected in fields such as medicine, transportation, and communication, enhancing quality of life.

Summary

In summary, the emergence of a methanol-centered industrial revolution requires:

i. Active government policy support and regulatory frameworks
ii. Balanced corporate strategies that align profitability with social responsibility
iii. Support for workforce adaptation and improved quality of life
iv. A shared vision for a sustainable and prosperous future

These elements must interact and evolve synergistically. Through collaboration and innovation, this revolution will progress—paving the way for a truly sustainable and prosperous methanol-based society.

9.6 Methanol Manufacturing Business: Opportunities and Key Players

Future Prospects for the Methanol Business

The methanol manufacturing business is expected to experience significant growth in the coming decades. In particular, large-and medium-scale methanol production plants will be prioritized to meet increasing demand. This section considers the supply chain needed to accelerate methanol production and the expected business opportunities (Table 9.5).

Table 9.5 Methanol production business

Industry	Example companies	Business opportunities
Methanol production technology	Carbon recycling technology startups, energy companies	Development of methanol production technology from CO_2, efficient utilization of renewable energy
Methanol fuel market	Oil companies, energy supply companies	Establishment of methanol fuel production and supply infrastructure, ensuring competitive fuel prices
Methanol-compatible engines and vehicles	Automobile manufacturers, engine manufacturers	Development of methanol-compatible engine technology, modification of existing engines to methanol compatibility
Chemical industry	Chemical manufacturers	Production of plastics, synthetic rubber, adhesives, solvents using methanol as a raw material
Methanol transport and storage	Logistics companies, tank manufacturers	Construction of safe methanol transport and storage systems
CO_2 capture and utilization	Environmental technology companies, industrial gas companies	CO_2 capture technology from factories and power plants, storage and transportation of CO_2
Infrastructure development	Construction companies, engineering companies	Construction of methanol production facilities, storage tanks, transportation pipelines
Policy and regulatory support	Consulting firms, regulatory authorities	Formulation and implementation support of policies promoting the methanol economy
Environmental monitoring and reporting	Environmental consultants, monitoring technology companies	Monitoring of CO_2 emissions, environmental impact assessment of methanol production

Development of Methanol Production Technologies

The advancement of methanol production technology is a critical focus area. Startups specializing in carbon recycling technology and energy companies engaged in CO_2 capture will play a key role. These entities are expected to drive innovation in methanol production from carbon dioxide and improve the efficient use of renewable energy.

Expansion of the Methanol Fuel Market

Expansion of the methanol fuel market is also anticipated. Oil companies and energy suppliers are likely to focus on developing methanol supply infrastructure and ensuring competitive fuel pricing.

Development of Methanol-Compatible Engines and Vehicles

The development of methanol-compatible engines and vehicles is another important area. Automobile manufacturers and engine manufacturers are expected to advance methanol fuel-compatible engine technologies and retrofit existing engines to use methanol.

Application of Methanol in the Chemical Industry

In the chemical industry, chemical manufacturers are anticipated to expand production of plastics, synthetic rubbers, adhesives, and solvents using methanol as a feedstock.

Safe Transportation and Storage of Methanol

Regarding the transportation and storage of methanol, logistics companies and tank manufacturers will be tasked with establishing safe and efficient systems to transport and store methanol.

CO_2 Capture and Utilization Technologies

For CO_2 capture and utilization, environmental technology companies and industrial gas companies are expected to develop CO_2 capture technologies for industrial facilities and power plants, as well as CO_2 storage and transportation solutions.

Infrastructure Development and Engineering

Construction firms and engineering companies will play a central role in constructing methanol production facilities, storage tanks, and transport pipelines, ensuring safe and efficient infrastructure to support the methanol economy.

Policy and Regulatory Support

In terms of policy and regulatory frameworks, consulting firms and regulatory authorities are expected to formulate and support policies that facilitate the growth of the methanol economy.

Environmental Monitoring and Assessment

Finally, environmental consultants and monitoring technology companies will be essential in conducting CO_2 emissions monitoring and performing environmental impact assessments for methanol production activities.

Socioeconomic Impact of the Methanol Economy

The methanol economy is expected to generate substantial socioeconomic benefits through job creation, regional revitalization, and industrial development. For example, a study in Scotland projected that carbon capture and utilization (CCU) projects could generate up to 4900 new jobs and contribute £490 million annually to the regional economy, highlighting the significant employment and economic potential of methanol-related infrastructure. Methanol production clusters are also expected to stimulate local economies. In regions where green methanol facilities have been launched, such as Iceland and coastal Spain, substantial economic activity has been reported in construction, logistics, and maintenance sectors.

While technological shifts will inevitably lead to some job transitions, coordinated retraining and reskilling programs can mitigate these effects. According to the EU Just Transition Mechanism reports, regions undergoing green energy transitions have maintained employment levels when vocational training is combined with targeted industrial investment. These findings underscore the importance of aligning methanol-related business development with social policies to ensure a just and inclusive energy transition.

9.7 Hydrogen Production and Renewable Energy Business

Importance of Hydrogen Production in a Methanol Society

In a methanol-centered energy system, developing and deploying hydrogen production technologies is a fundamental requirement. To meet the growing demand for synthetic methanol and other hydrogen-based fuels, it is critical to advance low-cost, high-efficiency hydrogen production processes. Here, we consider the hydrogen production business (Table 9.6) and the renewable energy business (Table 9.7).

Currently, green hydrogen—produced via electrolysis using renewable electricity—is considered the most promising long-term solution. However, it remains expensive, with delivered costs ranging from \$3 to \$6 per kilogram, where only about 15% accounts for production while the remaining 85% is attributed to storage, transportation, and refueling infrastructure. To address this, the U.S. Department of Energy has launched the Hydrogen Shot Initiative, targeting \$2/kg by 2026 and \$1/kg by 2031 as interim and final goals, respectively.

Table 9.6 Hydrogen production business

Industry	Example companies	Business opportunities
Hydrogen production technology	Hydrogen production equipment manufacturers, chemical industry companies	Advancements in electrolysis hydrogen production technology, development of low-cost, high-efficiency hydrogen production processes
Hydrogen supply infrastructure	Energy infrastructure companies, gas supply companies	Construction of hydrogen stations, establishment of pipeline networks, development of liquid hydrogen transportation systems
Hydrogen energy storage	Storage system manufacturers, energy management companies	Development and commercialization of high-efficiency hydrogen storage technologies
Hydrogen power generation technology	Generator manufacturers, fuel cell companies	Development and proliferation of hydrogen fuel cells, construction of hydrogen-powered generation plants

Table 9.7 Renewable energy business

Industry	Example companies	Business opportunities
Power generation technology	Solar power, wind power, hydropower technology companies	Development of efficient power generation technologies, reduction of generation costs, improvement of energy conversion efficiency
Energy supply stabilization	Energy storage companies, smart grid technology companies	Implementation of battery systems, flywheel energy storage, introduction of smart grids
Infrastructure development	Construction companies, infrastructure companies	Installation of wind turbines and solar panels, development of large-scale renewable energy farms
Energy management solutions	Energy management system companies, IoT technology companies	Efficient energy management and optimization, provision of energy consumption monitoring technologies
Financing clean energy projects	Financial institutions, investment funds	Investment in renewable energy and hydrogen-related projects, support for funding

To scale production, hydrogen production equipment manufacturers and chemical industry players must take the lead in driving down capital costs, enhancing system efficiencies, and promoting modular electrolyzer deployment.

Building the Methanol-Hydrogen Conversion Infrastructure

Establishing infrastructure to convert methanol into hydrogen—or vice versa—is another critical component. Methanol offers advantages as a liquid hydrogen carrier, simplifying storage and transport. Building this infrastructure requires cooperation among energy infrastructure companies, gas suppliers, and engineering firms.

Advancing Hydrogen Storage and Utilization

Hydrogen energy storage technologies are equally important. Storage system manufacturers and energy management companies are anticipated to develop and commercialize high-efficiency hydrogen storage systems. Additionally, the development of hydrogen-based power generation technologies is essential, with generator manufacturers and fuel cell companies expected to lead the development and deployment of hydrogen fuel cells and hydrogen-powered power plants.

Development of Power Generation Technologies

Regarding renewable energy-related industries, further advances in power generation technologies are required. Solar power, wind power, and hydropower technology companies are expected to focus on developing high-efficiency generation technologies, reducing generation costs, and improving energy conversion efficiencies.

Energy Supply Stabilization

To accommodate variable renewable sources and ensure reliability, investment in grid stabilization technologies is imperative. Companies specializing in battery storage, flywheels, and smart grid systems must deploy advanced energy management platforms. These platforms will enable real-time balancing of supply and demand, power quality optimization, and integrated control with hydrogen storage systems.

Infrastructure Development for Renewable Energy

Expanding renewable energy capacity will require large-scale infrastructure investments. Construction firms and infrastructure developers must take the lead in installing wind farms, solar parks, and transmission lines. Countries should adopt streamlined permitting systems to reduce lead times for project deployment.

Providing Energy Management Solutions

Energy efficiency and real-time control are key to maximizing system performance. Energy management system providers and IoT technology firms are expected to deliver AI-enabled solutions that optimize load distribution, minimize waste, and provide predictive maintenance across the energy network.

Financing Clean Energy Projects

Clean energy deployment at scale cannot occur without significant capital. Financial institutions and green investment funds play an essential role in mobilizing resources for hydrogen and renewable energy projects. According to the Hydrogen Council, cumulative global investment in clean hydrogen is expected to reach $680 billion by 2030. Governments must also offer public-private co-financing mechanisms, carbon pricing, and tax incentives to de-risk early-stage investments.

9.8 Business in the CO_2 Field: Opportunities and Challenges

The Urgency of CO_2 Removal

To mitigate global warming effectively, it is essential to prevent CO_2 from being emitted into the atmosphere and to remove the CO_2 that already exists in the air. As public awareness of climate change grows, the urgency to advance carbon capture, utilization, and storage (CCUS) is increasing. As of 2024, global CO_2 capture capacity stands at approximately 50 million tons per year, primarily driven by natural gas processing facilities that allow for low-cost, high-purity capture.

The International Energy Agency (IEA) projects that global capture capacity could reach 430 million tons annually by 2030, based on currently announced projects. However, considering annual CO_2 emissions are in the range of 35–45 billion tons, even capturing 1 billion tons—a frequently cited aspirational target—would only offset a small fraction of total emissions. To align with net-zero pathways, the scale of CO_2 removal must reach around 10 billion tons per year by mid-century.

Accelerating CO_2 removal requires building a supply chain. Table 9.8 shows promising CO_2-recyling businesses.

Table 9.8 CO_2 recycling business

Business type	Supply chain	Market size
Industrial CO_2 capture	Raw material supply: Factories, power plants	Capture of approximately 1 billion tons of CO_2 annually by 2030 (IEA)
	Capture technology: Absorption, adsorption, membrane separation	
	Storage facilities: Underground storage (CCS), compression tanks	
Synthetic fuel production	Hydrogen supply: Electrolysis hydrogen, fossil fuel-derived hydrogen	Approximately US$200 billion by 2050 (eFuel Alliance)
	Storage and transport: High-pressure gas cylinders, liquid hydrogen tanks, pipelines	
	Manufacturing facilities: Synthetic fuel plants	

CO_2 Capture Business: Current Status and Challenges

As of 2024, about 65% of CO_2 captured globally is derived from natural gas processing, where CO_2 can be separated relatively easily. In contrast, only 3% is captured from the steel sector, due to the technical complexity and high-temperature processes (e.g., blast furnaces and converters) that generate varied and diluted emissions. These processes demand significant investments in separation equipment—such as absorption towers, regeneration units, and compressors—making steel industry decarbonization through CCS particularly challenging. Consequently, alternative pathways such as hydrogen-based direct reduced iron (H_2-DRI) are gaining attention.

Emerging Markets and Complex Supply Chains

New CO_2 utilization markets are emerging, with increasingly complex supply chains. Industrial carbon capture involves collection from factories and power plants, followed by purification, compression, and either underground storage or conversion into products. According to the IEA and the Energy Transitions Commission, CCUS capacity has the potential to reach 1 billion tons per year by 2030, but only if deployment accelerates significantly.

CO_2-Derived Fuel Production

Captured CO_2 can be combined with hydrogen to produce synthetic fuels such as methanol and synthetic diesel. According to the eFuel Alliance, the synthetic fuel market is expected to reach US$200 billion by 2050. This supply chain includes hydrogen production (from electrolysis or fossil sources), storage and transport infrastructure (e.g., pipelines and pressure vessels), and synthesis plants.

CO_2-Derived Chemicals

CO_2 can be utilized to manufacture chemicals such as polycarbonates, by reacting with epoxy compounds. The global polycarbonate market is projected to reach US$25 billion by 2025. Similarly, urethane foams are synthesized using CO_2-derived polyols, with a market size estimated at US$90 billion by 2026. These supply chains require catalysts, pressurized reactors, and integrated chemical plant infrastructure.

Greenhouse Gas Fertilization

CO_2 can also be used to enhance plant growth in greenhouses, a practice known as greenhouse gas fertilization. This market is valued at US$2.5 billion by 2024. CO_2 is typically sourced from industrial facilities such as breweries or power plants, then captured, liquefied, stored, and distributed to agricultural sites.

CO_2 in the Beverage Industry

The carbonated beverage industry uses CO_2 to produce carbonated drinks. The market size is expected to reach US$413 billion by 2025. The supply chain involves CO_2 supply from breweries and power plants, capture and liquefaction technologies, liquefied CO_2 tanks, and beverage production facilities.

Economic and Environmental Implications

These projected market sizes reflect the expanding potential of CO_2-based industries to support both climate goals and economic development. While still insufficient to fully offset global emissions, the growth of the CO_2 business sector can serve as a vital component of a multi-pronged decarbonization strategy. A coordinated approach—combining investment, policy support, and technological innovation—will be essential to scale these solutions and realize their environmental and economic benefits.

Chapter 10
Technologies Required for a Methanol-Based Society

Abstract This chapter outlines a strategic approach for building a methanol society using existing, proven technologies, rather than waiting for entirely new innovations. Two technological pillars are emphasized: methanol production and energy utilization. To enable methanol's widespread adoption as a fuel and energy carrier, it is crucial to reduce the cost of hydrogen production and CO_2 recovery while concurrently expanding renewable energy capacity. In the short term, internal combustion engines (ICEs) offer a practical and mature method for methanol use, despite their modest efficiency. In the long term, fuel cells—especially when integrated with waste heat recovery—can achieve energy conversion efficiencies approaching 80%, making them essential for the future methanol economy. Heat utilization is highlighted as a vital factor in improving system efficiency. The deployment of low-cost, readily available heat management technologies should begin immediately, accompanied by ongoing technological refinement. This chapter concludes that by combining existing tools with well-planned integration of renewable energy, hydrogen, CO_2 capture, and thermal systems, a robust and scalable low-carbon energy infrastructure based on methanol can be realized.

Keywords Existing technologies · Hydrogen from biomass · Hydrogen from waste plastics · CO_2 recovery · Heat utilization · Formic acid · Internal combustion engines

10.1 Necessary Technologies in Reality

Conceptual Diagram and Energy Supply Chain

A conceptual diagram illustrating a feasible methanol energy supply chain is shown in Fig. 10.1. In this system, electricity is generated using biomass (such as wood), waste (such as waste plastics), and partially fossil resources. The exhaust gases from this process, primarily CO_2 and CO, are recovered and used as raw materials for methanol production. Technologies for thermal power generation using waste

J. Nakamura, M. Ishida, *Blueprint for a Methanol Society*,
https://doi.org/10.1007/978-981-95-3465-4_10

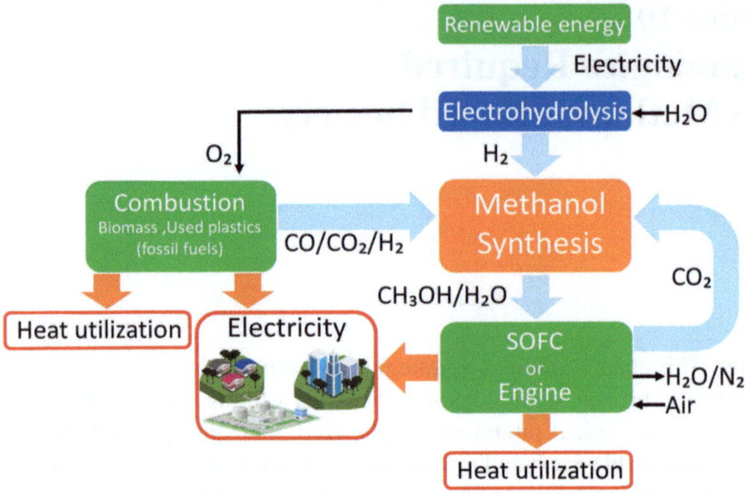

Fig. 10.1 Conceptual diagram of methanol energy supply chain

biomass and waste plastics have already been commercialized. While the foundational technologies already exist, targeted innovation is still needed to enhance performance and reduce costs at scale.

Integration of Renewable Energy and Electrolysis

In parallel, there will be significant introduction of renewable energy sources in the future. Electricity generated from renewable sources will be used for water electrolysis to obtain hydrogen (H_2) and oxygen (O_2). By using pure oxygen instead of air for thermal power generation, it eliminates the need to separate N_2 during combustion and allows for high-concentration CO_2 recovery. The produced H_2 serves as a raw material for methanol synthesis.

Utilization of Methanol

The synthesized methanol will be used in internal combustion engines and fuel cells (such as SOFC and PEFC) for power generation, as well as in automobiles, ships, and other transportation modes. In stationary applications, as much CO_2 as possible should be recovered.

Importance of Waste Heat Utilization

A critical focus of this system is on waste heat utilization. The key to a methanol-based society lies in establishing a high-efficiency system. Waste heat can be utilized for hot water supply and heating in facilities such as airports, hospitals, and shopping malls. It can also be applied to refrigeration using absorption and adsorption chillers and for drying, distillation, and chemical reactions in industrial processes. Waste heat can further be used for hydrogen production via methanol reforming, supporting hydrogen fuel cells. In some regions, waste heat can also be utilized for seawater desalination through distillation operations.

Improving Existing Technologies

The flow diagram in Fig. 10.1 demonstrates that this system can be implemented by enhancing existing technologies without relying on the development of entirely new technologies. Thus, improving current technologies to achieve high efficiency and low cost will be instrumental in promoting the construction of a methanol society. Delaying action until new technologies are developed would be too late to effectively address climate change.

Key Elements in the Methanol Energy Chain

The key components of this system include:

- Production of H_2, CO_2, and CO.
- Methanol synthesis.
- Use of methanol fuel cells (SOFC or PEFC).
- Methanol engines.
- Storage and transportation of methanol.
- Pipelines for CO_2 and H_2.
- Power generation technologies utilizing renewable energy.
- Waste heat utilization technologies.

The improvement and widespread deployment of these energy chain technologies are essential.

Realistic Approach to Hydrogen Production

Currently, the most cost-intensive aspect of the methanol society is hydrogen production [1]. Ideally, hydrogen would be produced via water electrolysis using electricity from renewable energy sources. However, this alone will not be sufficient to meet global demand in the near term. Therefore, it is more realistic to produce hydrogen through a combination of gasification of biomass and waste plastics. In the early stages, it will be inevitable to produce hydrogen and methanol from natural gas and coal to support the initial spread of methanol fuel.

10.2 Internal Combustion Engines: Methanol Engines

Applicability of Methanol in Conventional Engines

Even without employing advanced technologies such as solid oxide fuel cells (SOFCs), methanol can be readily utilized in internal combustion engines, which have long been established and have achieved a high level of technological maturity. Here, we propose an energy system that uses methanol as a fuel for an internal combustion engine and effectively utilizes waste heat, as shown in Fig. 10.2.

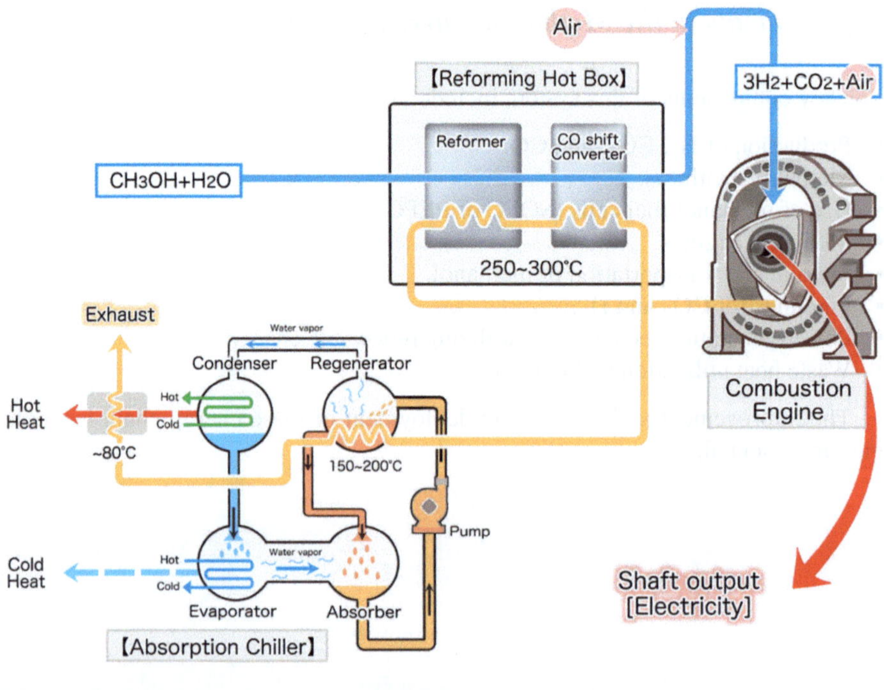

Fig. 10.2 Cascade energy use of methanol-water internal combustion engine

Advantages of Waste Heat Utilization

Although the energy conversion efficiency of internal combustion engines is relatively low, the significant amount of high-temperature exhaust heat generated can be effectively utilized. After serving as the heat source for the reforming reaction, the exhaust heat is subsequently utilized by an absorption chiller that provides both cooling and heating (with COP >1). Cascade heat utilization—the sequential use of waste heat at multiple temperature levels—can be considered a core principle of effective energy utilization.

Robust and Proven Technologies

Both the internal combustion engine and absorption refrigeration technologies are mature, having been refined and proven over many decades. These systems can generate electricity along with usable heating and cooling, making them reliable and practical energy solutions, particularly in times of emergency or disaster, contributing significantly to the maintenance of human life.

Handling and Environmental Considerations of Methanol

Methanol-water mixtures have the advantage of being relatively safe to handle. Depending on concentration, methanol-water mixtures may not be classified as hazardous materials under regulations such as the Japanese Fire Service Act. Lower-concentration solutions (<50 wt%) are generally considered safer and easier to handle. Small spills of methanol typically evaporate quickly and pose minimal immediate hazard; however, larger quantities or repeated exposure can be harmful to aquatic ecosystems and require appropriate management.

Practical Sufficiency Over Maximizing Efficiency

Even if maximum power generation efficiency is not pursued, methanol-fueled internal combustion engines, when integrated into systems with waste heat utilization, can sufficiently meet practical energy needs. In this form, they represent an effective and sustainable energy utilization strategy without the need for extensive technological innovations.

The Value of Proven and Mature Solutions

In this context, the immediate reorganization and application of existing mature technologies is crucial. While the potential of advanced technologies continues to grow, what is steadily required now is practical and actionable solutions that bring us closer to real-world implementation.

The Imperative of Embracing Change

The successful transition to a sustainable energy future requires the courage to embrace transformative change. The adoption of methanol as an energy carrier is steadily gaining global recognition, driven by its strong scientific foundation and economic feasibility. This growing momentum signifies a pivotal step toward realizing a methanol-based energy society on an international scale.

10.3 Formic Acid as an Energy Carrier

As repeatedly mentioned, the central idea is to convert renewable electricity into energy carriers such as hydrogen (H_2) and subsequently react H_2 with CO_2 to synthesize methanol. However, whether considering liquefied hydrogen or ammonia, a major challenge in building a hydrogen-based society is the high cost of hydrogen. Realistically, electrolysis of water remains the only technically viable method for producing hydrogen at scale, despite its significant costs in terms of transportation and storage of gaseous H_2. This is why formic acid emerges as a promising hydrogen carrier for the future [2]. Formic acid can be produced and stored directly using electrical energy, without first producing hydrogen gas as shown in Table 10.1. Specifically, it can be synthesized through the electrochemical reduction of CO_2:

$$CO_2 + H_2O \rightarrow HCOOH + \frac{1}{2}O_2 \tag{10.1}$$

The reduction reaction at neutral pH is expressed as:

$$CO_2 + H_2O + 2e^- \rightarrow HCOO^- + OH^- \quad -0.43\,\text{V vs. SHE} \quad \text{at pH} = 7.0, RT \tag{10.2}$$

Using electrodes made of materials such as Pb, In, Sn, and Cd, formic acid can be produced with 100% selectivity [3, 4].

Table 10.1 Various products from the electroreduction of CO_2 Adapted from [3] with the permission of Elsevier

Electrode	Potential(V) Vs.nhe	Current Density (mA cm^{-2})	Faradaic efficiency/%							
			CH_4	C_2H_4	EtOH	PrOH	CO	$HCOO^-$	H_2	Total
Cu	−1.44	5.0	33.3	25.5	5.7	3.0	1.3	9.4	20.5	103.5
Au	−1.14	5.0	0.0	0.0	0.0	0.0	87.1	0.7	10.2	98.0
Ag	−1.37	5.0	0.0	0.0	0.0	0.0	81.5	0.8	12.4	94.6
Zn	−1.54	5.0	0.0	0.0	0.0	0.0	79.4	6.1	9.9	95.4
Pd	−1.20	5.0	2.9	0.0	0.0	0.0	28.3	2.8	26.2	60.2
Ga	−1.24	5.0	0.0	0.0	0.0	0.0	23.2	0.0	79.0	102.0
Pb	−1.63	5.0	0.0	0.0	0.0	0.0	0.0	97.4	5.0	102.4
Hg	−1.51	0.5	0.0	0.0	0.0	0.0	0.0	99.5	0.0	99.5
In	−1.55	5.0	0.0	0.0	0.0	0.0	2.1	94.9	3.3	100.3
Sn	−1.48	5.0	0.0	0.0	0.0	0.0	7.1	88.4	4.6	100.1
Cd	−1.63	5.0	1.3	0.0	0.0	0.0	13.9	78.4	9.4	103.0
Tl	−1.60	5.0	0.0	0.0	0.0	0.0	0.0	95.1	6.2	101.3
Ni	−1.48	5.0	1.8	0.1	0.0	0.0	0.0	1.4	88.9	92.4
Fe	−0.91	5.0	0.0	0.0	0.0	0.0	0.0	0.0	94.8	94.8
Pt	−1.07	5.0	0.0	0.0	0.0	0.0	0.0	0.1	95.7	95.8
Ti	−1.60	5.0	0.0	0.0	0.0	0.0	Tr.	0.0	99.7	99.7

Electrolyte: 0.1 M $KHCO_3$, temperature: 18.5 + 0.5 °C

Advantages of Formic Acid as a Hydrogen Carrier

The attractiveness of this reaction lies in its simple elementary process: the direct addition of two hydrogen atoms to CO_2. Once $HCOO^-$ is formed, it thermally reacts with H_2O or H_3O^+ to produce formic acid (HCOOH). Because only one elementary step occurs at the electrode surface, if an appropriate catalyst is chosen, overpotential can be minimized, resulting in high energy conversion efficiency from electricity to formic acid. In contrast, the synthesis of methanol (CH_3OH) requires the addition of four hydrogens, doubling the complexity. Notably, formic acid can be thermally decomposed at relatively low temperatures (150–200 °C) to release hydrogen:

$$\mathrm{HCOOH} \rightarrow \mathrm{CO_2} + \mathrm{H_2}\ \mathrm{Cu\,or\,Ni\,catalyst}\ 150 \sim 200°C \tag{10.3}$$

Numerous studies have shown nearly 100% conversion rates and hydrogen selectivity above 90% at these temperatures [5].

Potential Applications and Practical Considerations

Formic acid, therefore, can act as a hydrogen carrier for fuel cells and internal combustion engines. For example, as shown in Fig. 10.3, formic acid is decomposed into CO_2 and H_2 through a thermal catalytic reactor, and the H_2 is supplied to a fuel cell. The recovered unreacted CO_2 is then electrochemically reacted with water to produce formic acid.

This setup can also be applied to stationary internal combustion engines in industrial complexes. In practice, formic acid production and decomposition equipment could be installed at industrial hubs (e.g., power plants, steel mills, and cement factories). Here, electricity from renewable sources would be used to produce formic acid from water and CO_2, creating an on-site energy storage and supply system. Waste heat from fuel cells or combustion can be reused for reforming reactions, further improving energy efficiency.

However, there are challenges. Due to its acidity, concentrated formic acid is corrosive and toxic, requiring appropriate materials such as fluororesins, polyethylene, or stainless steel and handling precautions, though dilute solutions are comparatively safer. Additionally, formic acid's lower energy density compared to methanol suggests it may be best used as a stationary energy storage medium rather than for long-distance transportation.

Finally, there is interest in methanol synthesis from formic acid. Although some reports describe direct methanol production from formic acid, the conversion

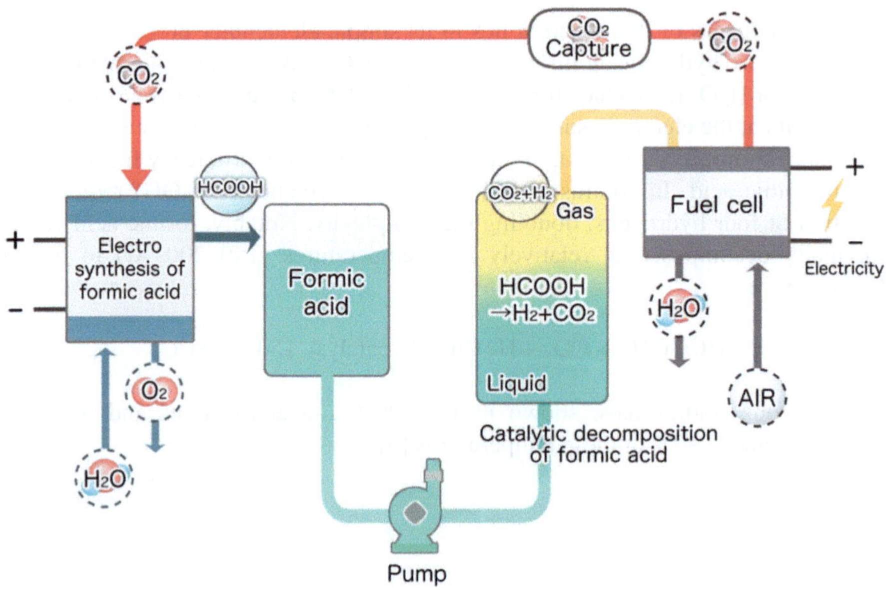

Fig. 10.3 Energy system that produces formic acid carriers directly from waste CO_2 and H_2O [7]

efficiency remains unclear [6]. An alternative would be to decompose formic acid back into CO_2 and H_2 and then synthesize methanol through known processes.

In summary, formic acid presents a viable, simpler, and direct electrochemical pathway for storing renewable electricity as a liquid energy carrier, bypassing some challenges of traditional hydrogen carriers. It also provides a practical and promising complement to methanol in building a sustainable, circular energy economy.

10.4 Promotion of Heat Energy Utilization

The Role of Heat Utilization in Future Energy Systems

A fundamental principle of energy use—beyond just renewable energy—is the effective utilization of heat. Historically, cheap fossil resources have enabled humanity to use energy freely. However, in a future where the use of fossil fuels is reduced, energy will inevitably become more expensive. The cost of renewable energy includes not only equipment but also the infrastructure for energy carriers, storage, and transportation. Recognizing and preparing for this reality is essential. Furthermore, expensive energy must not be wasted, which underscores the importance of heat utilization.

Heat Loss in Conventional Power Generation

When energy efficiency is low, virtually all lost energy is converted into heat. For example, consider today's coal-fired power plants:

- In conventional types, only about 35% of the energy from coal combustion is converted to electricity.
- In advanced modern plants, this has increased to about 50%.

Nonetheless, the remaining 50% of the energy becomes waste heat. In these plants, most of this heat is dissipated into the sea via cooling water. During the combustion of coal, the rotation of turbines, and the generation of electricity, significant waste heat is produced. This hot water—if harnessed as a usable energy source—could substantially improve the overall energy efficiency of thermal power generation.

Opportunities for Heat Utilization

Waste heat from thermal power plants can be used for:

- Heating and hot water supply in factories and buildings.
- Industrial process heat in various manufacturing sectors.

Such comprehensive heat management will become increasingly important as we transition away from fossil fuels.

Impact on CO_2 Emissions

Globally, the building sector accounts for approximately 9% of total CO_2 emissions (Fig. 2.3), primarily due to air conditioning, heating, and hot water in factories and buildings. Additionally, the industry sector uses significant amounts of electricity to produce heat for industrial processes. Therefore, by utilizing waste heat from power generation—whether in centralized or decentralized power sources—we can reduce CO_2 emissions. This heat recovery is essential for maximizing the conversion of fuels into electricity as a primary energy source, particularly in system power plants, but also increasingly in distributed power generation in the future.

In summary, effective heat energy utilization will be critical in a future methanol-based and renewable energy-powered society. By minimizing waste and maximizing energy efficiency, we can reduce environmental impact and build a more sustainable, carbon-neutral energy system.

10.5 Renewable Energy Technology and Hydrogen Production Technology

Advancement of Solar Power Generation

To realize a carbon-neutral society, large-scale deployment of renewable energy is indispensable. One of the most critical areas for development is photovoltaic (PV) technology.

- Currently, the conversion efficiency of commercial solar panels is about 20%.
- Ongoing research aims to achieve efficiencies of 25% or higher by 2025.
- In particular, perovskite solar cells have the theoretical potential to exceed 30% efficiency, and efforts toward practical applications are actively underway.

The levelized cost of electricity (LCOE) for solar power has decreased by about 80% over the past decade, reaching 3–6 cents per kWh as of today. The goal is to reduce LCOE further to 2–4 cents per kWh by 2030. Achieving this will require:

- Improving manufacturing process efficiency.
- Reducing material costs.
- Realizing economies of scale through mass production.

In addition, to address the variability inherent in solar power generation, the lithium-ion battery market is expected to grow at a compound annual growth rate (CAGR) of approximately 20% from 2020 to 2030, reaching a market size of about US$130 billion by 2030. This expansion will support the deployment of grid-scale energy storage systems and facilitate balancing power supply and demand.

Advancement of Wind Power Generation

Progress in wind power technology is also essential.

- Modern wind turbines typically have outputs of 2–3 MW per unit, while newer turbines exceed 12 MW.
- This improvement increases power generation efficiency and contributes to a more stable power supply.

As of 2020, the installed capacity of offshore wind power was about 29 GW, and this is expected to grow to over 200 GW by 2030. The development of floating wind turbines further enables installations in deep-water regions, expanding the potential for wind energy utilization.

The LCOE of wind power has fallen by approximately 50% over the past decade, currently standing at 5–10 cents per kWh, with a goal to reduce it to 3–6 cents per kWh by 2030. To achieve this, efforts must focus on:

- Mass production and standardization of turbines.
- Reduction of installation and maintenance costs.

Advancement of Energy Storage Technologies

Energy storage is indispensable for stabilizing renewable energy supply.

- The cost of lithium-ion batteries has dropped by about 85% over the past decade, reaching approximately US$137 per kWh in 2020.
- The target is to reduce this further to about US$70 per kWh by 2030.

Next-generation battery technologies such as solid-state batteries and lithium-sulfur batteries are projected to achieve 2–3 times the energy density of current batteries. Other technologies, including pumped hydroelectric storage, flywheel energy storage, and compressed air energy storage (CAES), also play vital roles in large-scale storage. For instance, pumped hydroelectric storage has an installed capacity of about 160 GW globally and serves as a major method of energy storage.

Smart Grid Integration and Demand Response

Smart grid technology is critical for integrating renewable energy effectively.

- Demand response systems are key to managing fluctuations in renewable supply.
- In the United States, the demand response market is expected to reach approximately US$5 billion annually by 2025.

The collaboration between smart meters and demand response systems enables real-time demand management. Moreover, integrating Distributed Energy Resources (DER)—projected to reach about 380 GW in the United States by 2025—enhances grid flexibility and promotes energy self-sufficiency.

Supporting the Methanol Society

For a methanol-based society, it is essential to:

i. Enhance the efficiency and cost-effectiveness of solar and wind power.
ii. Advance energy storage technologies.
iii. Deploy smart grid solutions.

The advancement of these technologies will enable a stable, low-cost supply of renewable energy, which will in turn lower the cost of methanol production and contribute to building a sustainable energy system.

In summary, the successful deployment of renewable energy technologies and energy storage systems—alongside smart grid integration—will provide the foundation for a carbon-neutral methanol-based energy system. Strategic investments in these areas are critical to ensure a sustainable future and mitigate climate change.

References

1. Schorn F, Breuer JL, Samsun RC, Schnorbus T, Heuser B, Peters R, Stolten D (2021) Methanol as a renewable energy carrier: an assessment of production and transportation costs for selected global locations. Adv Appl Energy 3:100050
2. Eppinger J, Huang KW (2017) Formic acid as a hydrogen energy carrier. ACS Energy Lett 2:188–195
3. Hori Y, Wakebe H, Tsukamoto T, Koga O (1994) Electrocatalytic process of CO selectivity in electrochemical reduction of CO_2 at metal electrodes in aqueous media. Electrochim Acta 39:1833–1839
4. Al-Tamreh SA, Ibrahim MH, El-Naas MH, Vaes J, Pant D, Benamor A, Amhamed A (2021) Electroreduction of carbon dioxide into Formate: a comprehensive review. ChemElectroChem 8:3207–3322

5. Faroldi BM, Conesa JM, Guerrero-Ruiz A, Rodríguez-Ramos I (2022) Efficient nickel and copper-based catalysts supported on modified graphite materials for the hydrogen production from formic acid decomposition. Appl Catal A 629:118419
6. Liu J, Zeng X, Cheng M, Yun J, Li Q, Jing Z, Jin F (2012) Reduction of formic acid to methanol under hydrothermal conditions in the presence of Cu and Zn. Bioresour Technol 114:658–662
7. Doyle A (2018) Formic acid-based fuel cell developed for hydrogen storage—News—The Chemical Engineer. https://www.thechemicalengineer.com/news/formic-acid-based-fuel-cell-developed-for-hydrogen-storage/. Accessed 31 May 2025

Chapter 11
Methanol Science and Technology

Abstract Methanol synthesis from CO_2 and H_2 via catalytic hydrogenation is already a commercialized and mature process. However, for methanol to serve as a practical energy carrier, further reductions in production cost and improvements in energy efficiency are essential. Key technological challenges include the development of low-cost, high-performance catalysts and the design of optimized reactors. In particular, Cu/ZnO catalysts remain the most widely used, and understanding their function is vital for designing next-generation systems. Recent advances in experimental and theoretical studies have significantly clarified the nature of active sites and reaction mechanisms, providing a foundation for targeted catalyst design. This chapter provides a detailed discussion of these findings, emphasizing the role of Cu-Zn synergy and surface intermediates. In addition, it addresses the importance of reactor engineering to manage the exothermic nature of the reaction and to effectively utilize the reaction heat. Shifting the equilibrium toward methanol production requires an integrated approach combining catalyst science, thermal management, and process optimization. Basic research, particularly that closely tied to industrial application, plays a central role in overcoming these barriers. This chapter concludes with a conceptual design of a cost-effective reactor system for improving methanol yield and energy efficiency.

Keywords Methanol synthesis from CO_2 · Cu/ZnO catalyst · Active sites · Reaction mechanism · Reactor design · Energy efficiency

11.1 Cost Reduction of Methanol Production

Overview and Key Challenge

In the context of methanol synthesis, the primary challenge is reducing the production cost. Establishing a low-cost methanol manufacturing process is central to building a methanol-based society. Achieving this requires dedicated efforts in

J. Nakamura, M. Ishida, *Blueprint for a Methanol Society*,
https://doi.org/10.1007/978-981-95-3465-4_11

science and technology. This section explores which manufacturing methods should be prioritized in the near term.

Biomass-Derived Methanol Synthesis Is Promising

As described in Sect. 7.4, there are several ways to produce methanol. There is direct synthesis from CO_2 emitted from combustion processes (direct synthesis by hydrogenation or electrolysis) or indirect synthesis from CO produced by gasification of coal or natural gas (conventional synthesis gas-based synthesis). And there is biomass-derived methanol synthesis. The National Renewable Energy Laboratory (NREL) has calculated the current and expected future costs of methanol production [1]. For commercially standard methanol production, it is 330 US dollars/ton, with raw material costs and operating costs accounting for a large proportion. The next most economical option is methanol production from biomass. It is currently estimated to be 390–500 US dollars/tonne, and 350–360 US dollars/ton in the future. The low capital cost is a feature of biomass utilization. Biomass utilization is generally considered to be carbon neutral and contributes to CO_2 reduction efforts. It is commercially feasible, so it should be tackled as a business opportunity immediately. However, resource limitations of biomass will likely delay its large-scale deployment.

Renewable Energy-Derived Methanol Synthesis

Renewable energy-derived H_2 is gaining global attention as a key to methanol production. While reducing the cost of renewable H_2 will take time, it is expected that methanol synthesis from H_2 produced via water electrolysis powered by renewable energy will expand. However, it will be more than 20 years before hydrogen derived from renewable energy enters a large-scale supply chain, given the current situation. This is because there are no large-scale synthetic chemical plants using electrochemical processes. Likewise, electrochemical methanol synthesis requires large capital investment, and NREL predicts that it will not fall below 1000 US dollars/tonne in the future.

A Realistic Hybrid Approach

A pragmatic proposal for the near term involves synthesizing methanol using natural gas, recovered CO_2, and renewable H_2. This can be achieved through the dry reforming of methane:

$$CH_4 + CO_2 \rightarrow 2CO + 2H_2 \tag{11.1}$$

Since this reaction does not match the stoichiometric ratio for methanol synthesis, an additional 2 moles of H_2 must be introduced:

$$CH_4 + CO_2 + 2H_2 \rightarrow 2CO + 4H_2 \rightarrow 2CH_3OH \tag{11.2}$$

This means 50% of the methanol is produced using renewable H_2. Even when natural gas is involved, expanding the methanol value chain is crucial.

Long-Term Vision and Interim Measures

Ultimately, it is desirable to integrate synthesis gas derived from biomass and fossil fuels with renewable H_2 to establish a robust supply chain. In the long term, direct methanol synthesis thermally or electrochemically from emitted CO_2 and renewable H_2 should become the mainstream method, aligned with carbon recycling principles. However, to accelerate the realization of a methanol-based society, it is essential to pursue cost reductions across all manufacturing methods in the interim and build systems that support this transition.

11.2 Selection of Inexpensive Catalysts

The Central Role of Catalysts in a Methanol Society

In a methanol-based society, chemical reactions play a pivotal role. Specifically, CO_2 must be chemically converted with hydrogen derived from water to produce methanol. As a result, chemistry will increasingly become a key discipline in the energy sector. Traditionally, the chemical industry has focused on producing plastics, pharmaceuticals, food, and resins from petroleum, with most products discarded after use. In the upcoming circular economy, however, chemical reactions will be critical not just in the chemical industry but also in the energy industry, particularly for producing artificial fuels. Covering the entire energy demand of humankind solely with nuclear power and batteries is not feasible. This transition represents a significant energy and industrial revolution, requiring a fresh perspective on artificial fuel production beyond conventional approaches in the chemical industry.

Catalyst Demand and Challenges

In practical chemical reactions, catalysts are indispensable. Looking ahead, with catalysts being employed from the chemical industry to the energy sector, the demand for catalysts will skyrocket. However, many of the most active catalysts are based on precious metals, such as platinum (Pt), which is widely used in the petrochemical industry. For example, Pt is essential in the reforming of light hydrocarbons (like naphtha) to produce high-octane gasoline and aromatics, in fluid catalytic cracking (FCC) for converting heavy oil into lighter hydrocarbons, and in hydrosilylation reactions for producing silicone resins. In these processes, it is estimated that several tens to hundreds of tons of Pt are used globally.

Pt in Fuel Cell Vehicles: A Bottleneck

Pt catalysts are also employed in proton exchange membrane fuel cells (PEMFCs). Consider the amount of Pt used in hydrogen fuel cell vehicles: currently, 15–30 g of Pt is needed per vehicle. With around 1.5 billion vehicles worldwide, if 10% of these were fuel cell vehicles, 3000 tons of Pt would be required. Given that the annual Pt production is approximately 180 tons, and that existing Pt stockpiles are limited to several thousand tons, relying on Pt catalysts precludes widespread adoption of fuel cell vehicles at present. The high durability of Pt under the acidic and low-temperature conditions of PEMFCs (80–100 °C) is a major reason for its use. Many other metals dissolve and lack the required durability in such environments.

Emerging Alternatives: Nitrogen-Doped Carbon Catalysts

Given these limitations, nitrogen-doped carbon materials are gaining attention as promising Pt alternatives for PEMFC cathode catalysts [2]. In PEMFCs, the key reactions are:

- Anode reaction (hydrogen oxidation reaction, HOR):

$$2H_2 \rightarrow 4H^+ + 4e^- \text{ Anode reaction} \tag{11.3}$$

- Cathode reaction (oxygen reduction reaction, ORR):

$$O_2 + 4H^+ + 4e^- \rightarrow 2H_2O \text{ Cathode reaction} \tag{11.4}$$

Currently, Pt-based catalysts dominate both reactions, but cathode catalysts account for 70–90% of total Pt usage because the ORR is slower than the HOR and thus requires more catalyst to accelerate the reaction rate. In contrast, the anode reaction is relatively simple, involving only H–H bond cleavage. The cathode reaction, however, includes O–O bond cleavage and four proton attacks on oxygen atoms, encompassing five elementary steps (excluding adsorption/desorption), making it much slower and thus more catalyst-intensive. Overvoltage arises from slow elementary steps, representing energy losses.

Broader Impact of Carbon Catalysts

Carbon-based catalysts also show promise for hydrogen production by water splitting and are increasingly being investigated [3, 4]. Figure 11.1 highlights the growing research focus on carbon electrode catalysts, which are affordable and conductive. In a carbon-neutral society, carbon catalysts are an essential research frontier.

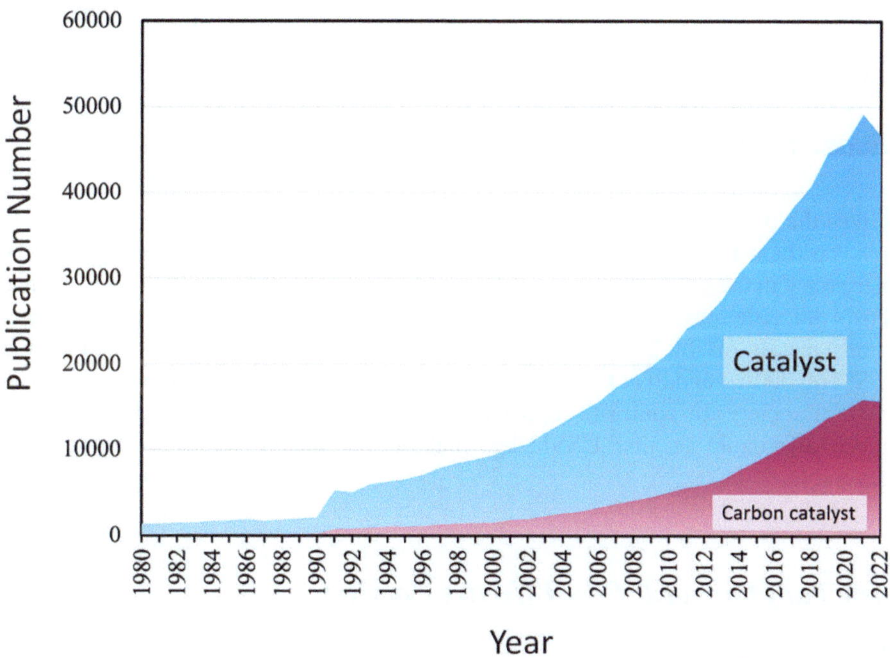

Fig. 11.1 The ratio of papers with the keyword "carbon catalyst" to the total number of papers with the keyword "catalyst"

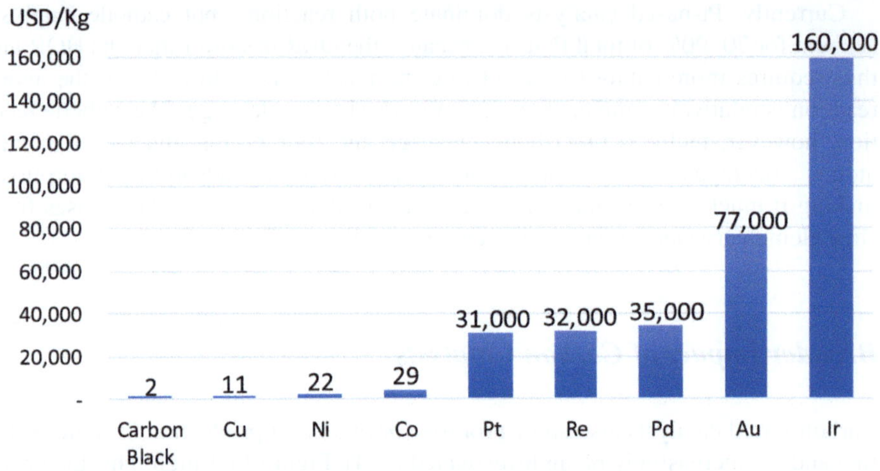

※ Cu	https://www.hashimoto-kosan.jp/copper/ （2024.5）
Ni,Co,Au	https://www.asaka.co.jp/undertaking/metals/price/(2024.5)
Carbon Black	https://www.analystjapan.com/Pricing-data/carbon-black-30(2024.3)
Re	https://www.sigmaaldrich.com/JP/ja/product/aldrich/267279(2024.5)
Pt,Pd,Ir	https://tanaka-preciousmetals.com/jp/library/rate/ （2024.4）

Fig. 11.2 Metal prices around May 2024 (US$/kg)

Economic Imperative: Moving Away from Precious Metals

Given the prohibitive cost and limited supply of precious metals, it is unrealistic to rely on them for future energy applications. While some precious metals may be necessary in the early phases of methanol and fuel cell adoption, their reserves and costs preclude long-term reliance. Figure 11.2 illustrates the market prices of precious metals (as of 2024): Iridium (Ir), used in water electrolysis, is priced at US$160,000/kg and Pt at US$31,000/kg. These costs are major barriers to a carbon-neutral society. In contrast, copper (Cu)—used in methanol synthesis—is much more affordable at just US$11/kg, making methanol synthesis economically favorable.

The Path Forward: Cost-Effective Catalysts

To enable industrial-scale methanol synthesis and the broader hydrogen economy, future catalyst research must focus on inexpensive, abundant materials. The transition to a carbon-neutral society hinges on the development of low-cost, high-performance catalysts—a challenge that is at the heart of this energy and industrial revolution.

11.3 Methanol Synthesis from CO and CO_2

Diversity of Feedstocks

Methanol synthesis is notable for its flexibility: it can be produced from either CO or CO_2. This versatility opens up a range of feedstock options. For example, when biomass is gasified, the resulting syngas contains a mixture of CO, CO_2, and H_2, which can be used directly as a reactant without separation. Here, we focus on the differences in catalysts, reaction conditions, and reaction mechanisms in methanol synthesis from CO and CO_2 to optimize production according to the feedstock composition. The basic reactions are:

From CO:

$$CO + 2H_2 \rightleftharpoons CH_3OH \quad \Delta_r H^\circ = -91.0\,\mathrm{kJ/mol} \tag{11.5}$$

From CO_2:

$$CO_2 + 3H_2 \rightleftharpoons CH_3OH + H_2O \quad \Delta_r H^\circ = -49.4\,\mathrm{kJ/mol} \tag{11.6}$$

As discussed in Sect. 7.5, the equilibrium conversion for methanol synthesis from CO is higher than that from CO_2 at lower temperatures due to the larger exothermic heat release in the former reaction. The standard reaction conditions are around 250 °C and 50–100 atm.

Heat Management in Reactor Design

A key difference in reactor design between the two processes is the need for heat removal. CO hydrogenation releases more heat and thus requires more intensive heat removal systems, making efficient heat recovery critical for artificial fuel production. In current methanol production, heat recovery is not extensively implemented because fossil fuels have historically been inexpensive compared to the costs of heat recovery systems. However, in the context of methanol as an energy carrier, this heat represents an important portion of the energy and should not be wasted. Facilities for heat recovery should be incorporated into methanol synthesis plants, with systems for hot water supply as part of an integrated approach.

Catalyst Stability and Degradation

Although Cu/ZnO-based catalysts are used for both CO and CO_2 hydrogenation, the chemical environments differ. The reactant mixture for CO hydrogenation is highly reducing, whereas CO_2 hydrogenation involves an oxidizing atmosphere.

Catalyst degradation also differs: for CO hydrogenation, prolonged operation can lead to ZnO reduction and catalyst sintering. In CO_2 hydrogenation, catalyst degradation is linked to water formation. Interestingly, when CO_2 and H_2O are present in the gas phase during CO hydrogenation, this mixture can help mitigate catalyst degradation.

Reverse Water-Gas Shift Reaction and Reaction Pathways

In CO_2 hydrogenation, the reverse water-gas shift (RWGS) reaction must be considered:

$$CO_2 + H_2 \rightleftharpoons CO + H_2O \quad \Delta_r H^\circ = +41.6\,\text{kJ/mol} \tag{11.7}$$

A key question is whether methanol is directly synthesized from CO_2 or if CO is first formed via RWGS and then converted to methanol. Isotope tracer studies (e.g., using ^{14}CO and $^{14}CO_2$) from the 1980s revealed that methanol formation from CO_2 is significantly faster than from CO. Figure 11.3 shows the concentration of ^{14}C in methanol under reaction conditions of 523 K and 50 atm using a $Cu/ZnO/Al_2O_3$ catalyst. As the space velocity increases, the ^{14}C concentration in methanol closely matches that in CO_2, confirming that methanol is produced directly from CO_2.

Other studies have supported these findings [4–6], showing that while methanol can be produced from both CO and CO_2, it proceeds several times to 20 times faster from CO_2, depending on the catalyst and reaction conditions. Table 11.1 presents turnover frequency (TOF) data for CO and methanol production during CO_2 hydrogenation over Cu(111), Zn/Cu(111), Cu/SiO_2, and Cu/ZnO catalysts [7]. The TOF for methanol production is $0.2–2.7 \times 10^{-2}$ molecules s^{-1} $site^{-1}$, while the TOF for CO production is 1.1–90 molecules s^{-1} $site^{-1}$, indicating that the RWGS reaction is significantly faster than methanol synthesis.

Equilibrium Limitations and Low-Temperature Challenges

Although methanol production from CO_2 is faster, practical yields are limited by equilibrium constraints. Because methanol synthesis from CO_2 also produces equimolar H_2O, this requires additional separation steps. More critically, at the same reaction temperature, the equilibrium yield of methanol from CO_2 hydrogenation is much lower than from CO hydrogenation. For example, achieving a 90% methanol yield from CO_2 hydrogenation at 10 MPa would require a reaction temperature as low as 320 K—far below current industrial practice.

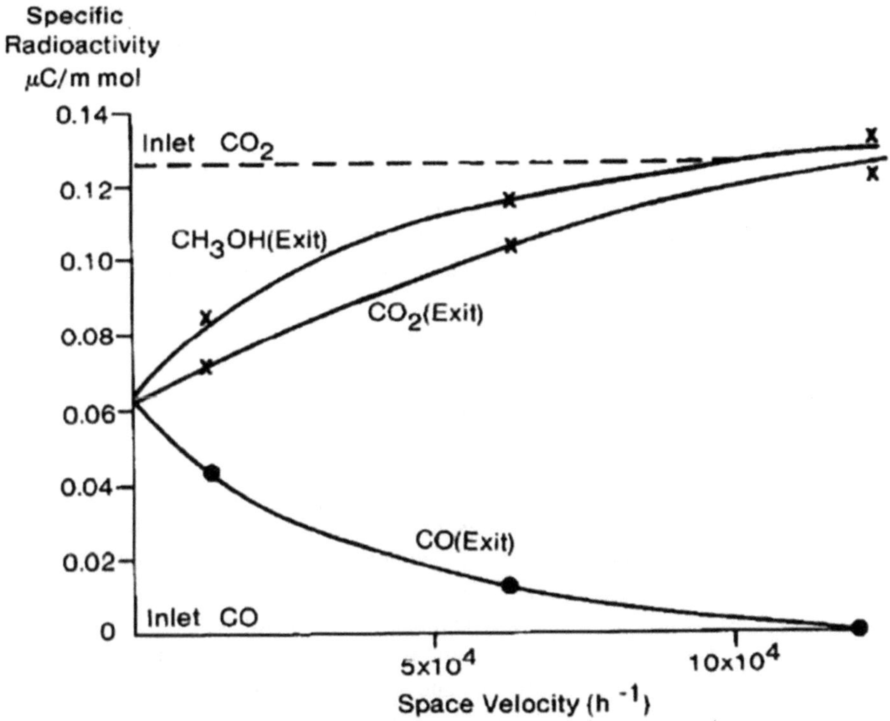

Fig. 11.3 Relationship between the specific radioactivity (radioactivity per unit mole) corresponding to the ^{14}C concentration in CH_3OH, CO_2, and CO and the space velocity in a reaction using 10%CO/10%CO_2/80%H_2 containing $^{14}CO_2$. Cu/ZnO/Al_2O_3 catalyst, reaction temperature 523 K, 50 atm. (Reprinted from [3] with the permission of Elsevier)

Table 11.1 Comparison of the reaction rates of various catalysts for methanol synthesis and reverse water-gas shift reaction (TOF). Adapted from [7] with the permission of Springer Nature

	Cu(111)	Zn/Cu(111)	Cu/SiO_2	Cu/ZnO
Temperature (K)	523	523	523	523
Pressure[a] (atm)	18	18	20	18
Feed rate or SV	5.56 cm^3/min (at 18 atm)	5.56 cm^3/min (at 18 atm)	1.8×10^5 h^{-1} (at 1 atm)	1.2×10^5 h^{-1} (at 1 atm)
TOF(CH_3OH)[b]	0.2×10^{-2}	2.7×10^{-2}	0.8×10^{-2}	1.7×10^{-2}
TOF(CO)[b]	90	70	5.65	1.12

[a] $H_2/CO_2 = 3$
[b] TOF (molecules/site/s)

The Cu/ZnO catalyst has been the most studied system, and while Pd catalysts have shown some promise, neither has yet reached the performance needed for low-temperature operation. Recent studies have identified catalysts with some low-temperature activity, but large-scale CO_2 hydrogenation still relies on the established Cu catalysts due to their cost-effectiveness and availability.

In summary, while methanol synthesis from CO_2 benefits from faster kinetics compared to CO, practical challenges—including equilibrium limitations and catalyst stability—still need to be addressed. The development of more active, low-cost catalysts remains an important research frontier for building a robust methanol economy.

11.4 Function of Cu/ZnO Catalysts

To prepare highly active Cu/ZnO catalysts for methanol synthesis, it is crucial to understand the nature of the active sites and how they evolve under different reaction conditions. One of the complexities in methanol synthesis is that the active sites differ depending on whether CO or CO_2 is used as the reactant. Furthermore, in both cases, there is not just a single type of active site involved.

In summary, both Cu–Zn alloy sites and Cu sites are essential for methanol synthesis from CO_2, while Cu–O–Zn sites and Cu sites are needed for methanol synthesis from CO. The common feature is that Cu sites are required for hydrogen dissociation; however, the active sites for the subsequent hydrogenation steps differ.

Experimental Evidence of Active Sites Using a Physical Mixture of Cu/SiO$_2$ and ZnO/SiO$_2$

Nakamura/Fujitani group has demonstrated this difference in catalytic active sites using a physical mixture of Cu/SiO_2 and ZnO/SiO_2. In these experiments, the catalyst components of Cu and ZnO were individually supported on the inert support SiO_2. Figure 11.4 shows the methanol and CO production from CO_2 hydrogenation after reduction with H_2 at various temperatures using this physically mixed catalyst [6]. The (Zn)Cu/SiO_2 catalyst in the △ plot was prepared by reducing Cu/SiO_2 and ZnO/SiO_2 to allow Zn migration from ZnO to Cu and CuZn alloy formation, followed by sieving to ensure well-mixed active sites.

As the reduction temperature increases, both the physically mixed catalyst and the (Zn)Cu/SiO_2 catalyst exhibit enhanced methanol synthesis activity. This enhancement is attributed to the migration of Zn species from ZnO to Cu particles, forming Cu–Zn alloy (brass) active sites. XRD measurements show that the Zn dissolved in Cu particles reaches a saturation limit of approximately 20 atom% at 723 K.

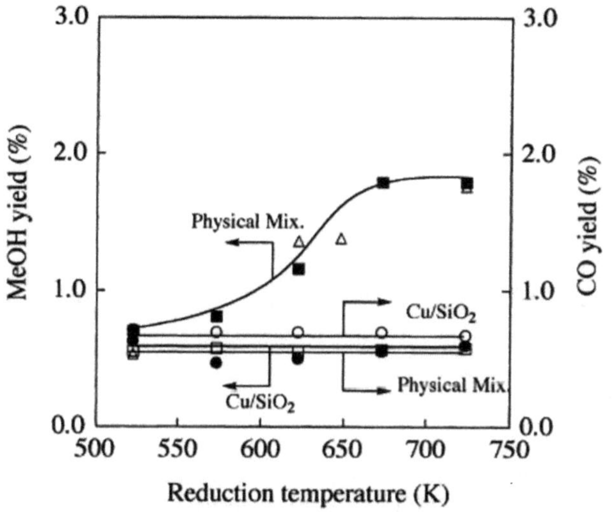

Fig. 11.4 Relationship between methanol and CO yield and hydrogen reduction temperature in $CO_2/3H_2$ using Cu/SiO_2 catalyst and Cu/SiO_2 + ZnO/SiO_2 physical mixed catalyst, fixed-bed flow reactor, 523 K, 50 atm, △ is (Zn)Cu/SiO_2 catalyst (Cu/SiO_2 catalyst with migrated Zn). (Reprinted from [17] with the permission of Elsevier)

Conversely, the amount of CO produced remains independent of the reduction temperature and matches that of pure Cu/SiO_2 catalysts with no Zn. This observation confirms that the reverse water-gas shift reaction occurs exclusively on the Cu surface.

Activation Energy Analysis and Active Site Characteristics

Further analysis of the apparent activation energies for methanol and CO production across different reduction temperatures (Fig. 11.5) shows that the quality of active sites remains constant, while the number of active sites increases. These results underscore the importance of Zn migration to Cu for forming methanol-synthesis-active sites, while the properties of the Cu surface responsible for the reverse water-gas shift reaction remain unchanged.

Distinction of Active Sites for CO_2 and CO Hydrogenation

Figure 11.6 plots the methanol yield from CO_2 and CO hydrogenation at different reduction temperatures. With increasing reduction temperature, CO_2 hydrogenation activity rises, while CO hydrogenation activity decreases. This clear divergence

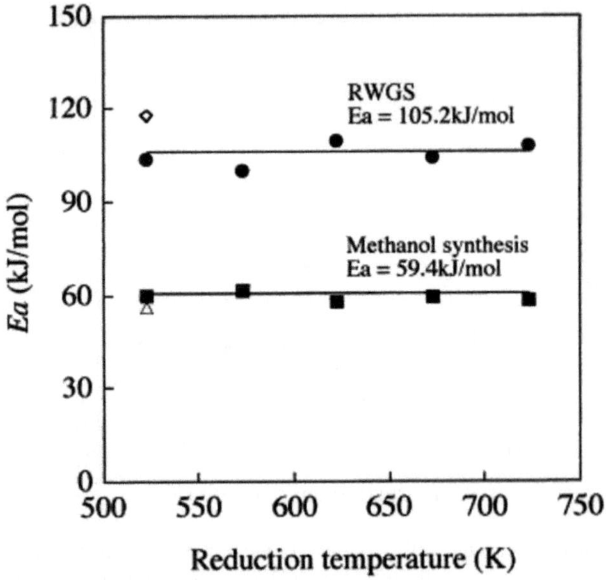

Fig. 11.5 Relationship between activation energy and hydrogen reduction temperature for methanol synthesis from $CO_2/3H_2$ and reverse water-gas shift reaction, catalyst: $Cu/SiO_2 + ZnO/SiO_2$ physical mixed catalyst, fixed-bed flow reactor, reaction temperature: 523 K, 50 atm, △, ◇ are $Cu/ZnO/Al_2O_3$ catalyst. (Reprinted from [17] with the permission of Elsevier)

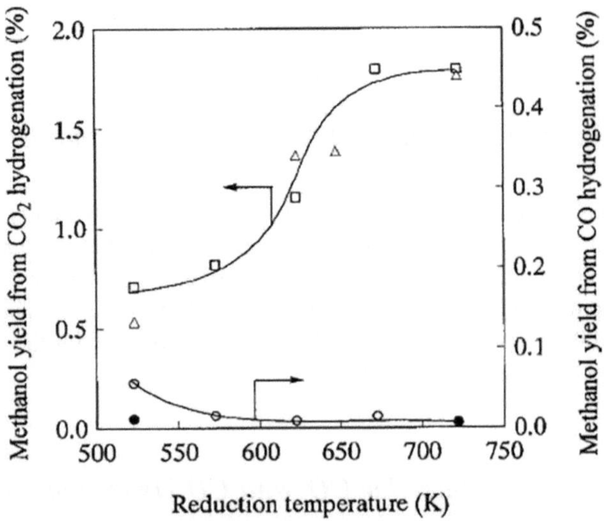

Fig. 11.6 Relationship between methanol synthesis yield from CO_2 and CO and hydrogen reduction temperature, catalyst: $Cu/SiO_2 + ZnO/SiO_2$, fixed-bed flow reactor, reaction temperature 523 K, 50 atm ($H_2/CO = 2$, $H_2/CO_2 = 3$) ●: CO hydrogenation with Cu/SiO_2 catalyst, △: CO_2 hydrogenation with $(Zn)Cu/SiO_2$ catalyst. (Reprinted from [6] with the permission of Springer Nature)

demonstrates that the Cu–Zn alloy sites, effective for CO_2 hydrogenation, are inactive for CO hydrogenation (see Fig. 11.9).

Influence of Oxidation on Active Site Formation

To confirm the role of oxidation in forming active sites for CO hydrogenation, the (Zn)Cu/SiO_2 catalyst was oxidized at 673 K. Figure 11.7 shows a significant increase in methanol synthesis activity after oxidation. Subsequent reduction at 723 K sharply reduces activity, confirming that Cu–O–Zn sites in the oxidized state are essential for CO hydrogenation, alongside Cu sites for hydrogen dissociation.

Determination of Reduction Temperature Threshold

To further investigate, we fixed the oxidation temperature of (Zn)/Cu/SiO_2 at 623 K and varied the reduction temperature of the physically mixed Cu/SiO_2 + ZnO/SiO_2 catalyst. As shown in Fig. 11.8, activity increases with higher reduction temperatures. The activation energies for methanol synthesis across samples prepared at

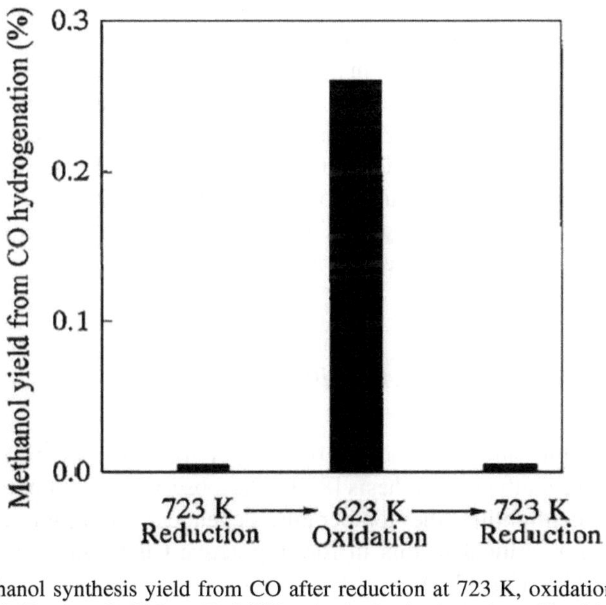

Fig. 11.7 Methanol synthesis yield from CO after reduction at 723 K, oxidation at 623 K, and reduction at 723 K using Cu/SiO_2 + ZnO/SiO_2 physical mixed catalyst. Fixed-bed flow reactor, reaction temperature 523 K, 50 atm (H_2/CO = 2). (Reprinted from [6] with the permission of Springer Nature)

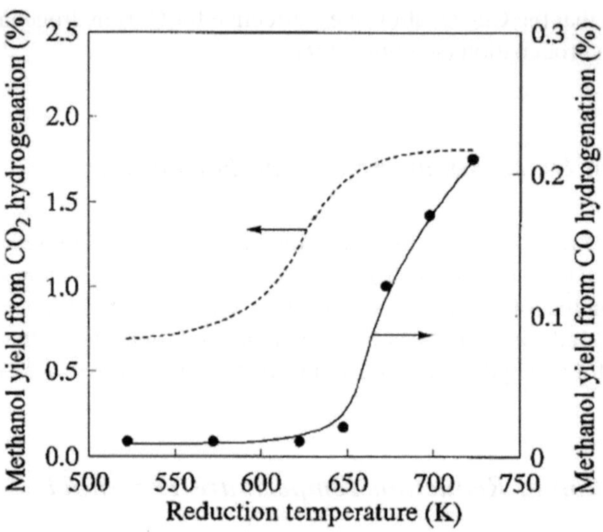

Fig. 11.8 The yield of methanol synthesis from CO after reduction of $Cu/SiO_2 + ZnO/SiO_2$ physical mixed catalyst at various temperatures from 523 K to 723 K, followed by oxidation at 623 K, in a fixed-bed flow reactor, at 523 K and 50 atm ($H_2/CO = 2$). The dashed line shows the hydrogenation of CO_2 by $(Zn)Cu/SiO_2$ catalyst . (Reprinted from [6] with the permission of Springer Nature)

reduction temperatures of 673 K, 698 K, and 723 K remain consistent (86–96 kJ/mol), indicating that higher reduction temperatures increase the number of Cu–O–Zn active sites without altering their intrinsic activity.

Interestingly, a sudden increase in activity occurs above 650 K, distinct from the behavior of CO_2 hydrogenation (dashed line in Fig. 11.8). XRD analysis of Cu/SiO_2 reduced at 723 K shows that the Zn content reaches 13–15%, close to the solubility limit (~20%). Furthermore, studies with Zn/Cu(111) model catalysts suggest that below 20% Zn content, Zn is reduced to metallic Zn under CO_2/H_2, but above 20%, ZnO becomes stabilized. This suggests that when Zn content exceeds the solubility limit, excess Zn is readily oxidized, forming Cu–O–Zn active sites.

Implications for Catalytic Activity

These results align with prior findings by King et al., which identified Cu^+ species as active sites for methanol synthesis [8, 9]. Our observations support this interpretation while highlighting the temperature-dependent migration and oxidation behavior of Zn as critical factors in forming active Cu–O–Zn sites for methanol synthesis from CO (Fig. 11.9).

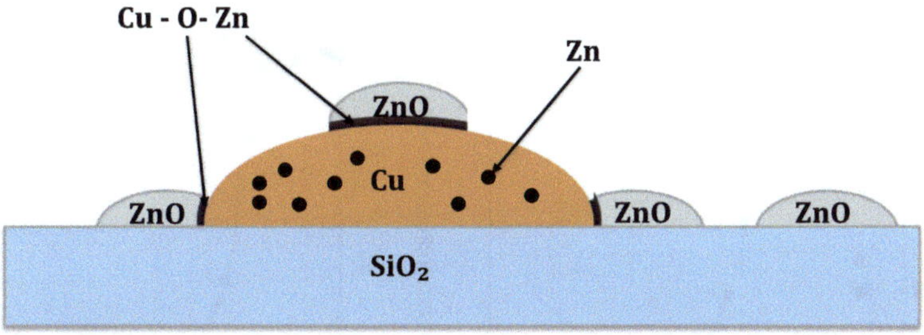

Fig. 11.9 Image of (Zn)Cu/SiO_2 catalyst. It was concluded that the Cu-Zn sites are the active sites for methanol synthesis by CO_2 hydrogenation, and the Cu-O-Zn sites are the active sites for methanol synthesis by CO hydrogenation

11.5 Mechanism and Active Sites in Methanol Synthesis with Cu/ZnO Catalysts

Zn/Cu(111) Model Catalyst Studies

Nakamura, Fujitani, and colleagues have extensively studied model catalysts to confirm their hypotheses regarding active sites for methanol synthesis, based on observations from powdered catalysts. As discussed in Sect. 11.4, experiments using physically mixed catalysts suggested that Zn species migrate from ZnO particles to the surfaces of Cu particles during high-temperature hydrogen reduction, forming Cu–Zn alloy active sites. To validate this, they prepared a Zn/Cu(111) model catalyst by depositing Zn atoms onto a clean Cu(111) surface under vacuum and investigated its catalytic activity for methanol synthesis via CO_2 hydrogenation, as well as for the reverse water-gas shift reaction. The experiments were notable for coupling a high-pressure reactor to an ultra-high vacuum XPS system.

Agreement of Activity Between Powder Catalyst and Model Catalyst

Figures 11.10a, b, displays the turnover frequency (TOF) for CH_3OH and CO production, respectively, under a high-pressure reaction environment (CO_2:H_2 = 1:3, 18 atm, 523 K) [10]. The results demonstrated that Zn deposition significantly enhances methanol synthesis activity, while CO production activity gradually decreases. This aligns well with powdered catalyst experiments, confirming that Cu–Zn alloy sites are responsible for methanol synthesis, while CO is formed at Cu surface sites

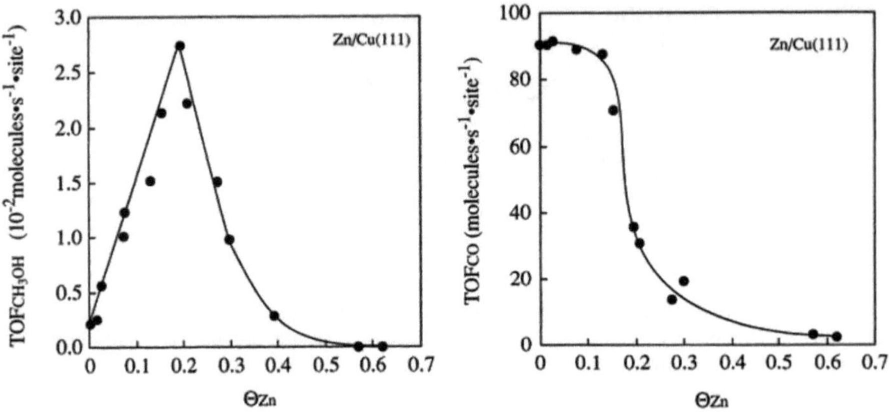

Fig. 11.10 TOF of CH_3OH (**a**) and CO (**b**) formation on a Cu(111) model catalyst with Zn deposition. $P(H_2) = 13.5$ atm, $P(CO_2) = 4.5$ atm, 523 K. (Reprinted from [10] with the permission of Elsevier)

The horizontal axis in these figures, Θ_{Zn}, represents the coverage of Zn on the Cu(111) surface, where $\Theta_{Zn} = 1$ corresponds to the number of Cu surface atoms (1.77×10^{15} atoms/cm^2). At $\Theta_{Zn} = 0.19$, methanol production reaches its maximum TOF (2.7×10^{-2} molecules·s^{-1}·site^{-1}) and an activation energy of 83.7 kJ/mol—values consistent with those for the Cu/ZnO catalyst under the same conditions (1.6×10^{-2} molecules·s^{-1}·site^{-1} and 69.3 kJ/mol, respectively). These findings validate the use of Zn/Cu(111) as an effective model catalyst for mechanistic studies.

Following these experiments, the research team has closely examined the reaction mechanism, reaction dynamics, and electronic structure using Zn/Cu(111) with $\Theta_{Zn} = 0.19$ as a model catalyst. Establishing such a reliable model system enables a unique opportunity: the catalyst's mechanism can be elucidated at the atomic and electronic level using comprehensive surface science methods.

Evidence that Cu–Zn Alloy Sites Are Active Sites

A notable observation is that the TOF increases linearly with Θ_{Zn} up to 0.19 and then sharply declines at higher coverages. To understand this, ex situ XPS measurements were performed after CO_2 hydrogenation. Following the reaction at 523 K under high pressure, the sample was cooled to 373 K in the reaction gas atmosphere, then evacuated and analyzed by XPS. Careful evacuation at 373 K was necessary to prevent decomposition of intermediate species such as formate ($HCOO_a$).

Figure 11.11 shows the O 1 s spectrum analysis of the Zn/Cu(111) surface after CO_2 hydrogenation at $\Theta_{Zn} = 0.19$, revealing substantial changes. At and below

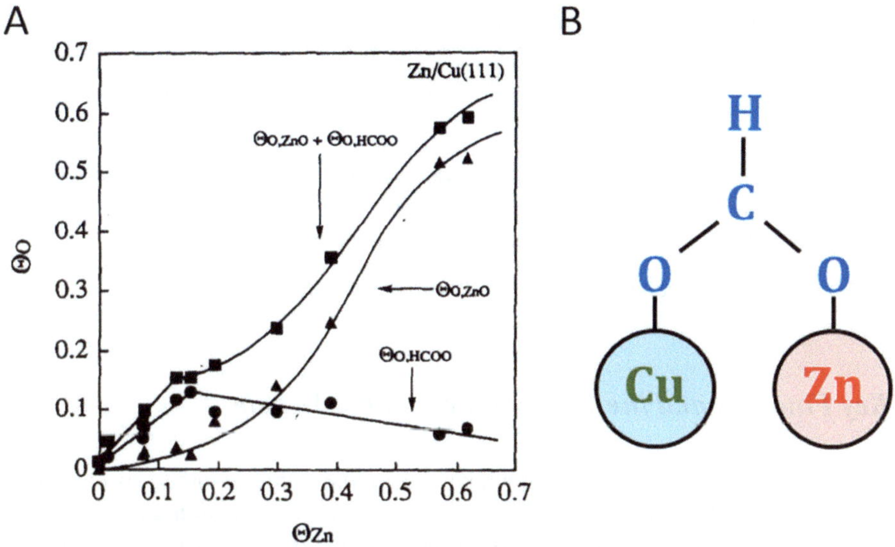

Fig. 11.11 Ex situ XPS analysis of the methanol synthesis reaction on a Zn/Cu(111) model catalyst . The oxygen coverage from ZnO, the oxygen coverage from formate, and the sum of both coverages are shown as a function of Zn coverage. (Reprinted from [10] with the permission of Elsevier)

Θ_{Zn} = 0.19, Zn remains metallic, while above this coverage, Zn is oxidized to ZnO. As the quantity of Cu–Zn alloy sites increases, methanol synthesis activity rises; however, the formation of ZnO leads to a significant drop in activity. These findings confirm that Cu–Zn alloy sites are the active centers for methanol synthesis from CO_2, consistent with results from powdered catalysts.

The Ongoing Debate: Active Sites in Methanol Synthesis

Let us now discuss the active site controversy in the journal *Science*, in which our group has been involved [11–13]. As previously mentioned, Nakamura and Fujitani have proposed—and it has been widely accepted—that the active site for methanol synthesis is the "Cu–Zn alloy model." However, in 2017, a paper was published in *Science* claiming that the active sites are actually formed at the interface between Cu and ZnO [11]. Interestingly, Nakamura and Fujitani had themselves reported similar findings (specifically, ZnO formation) prior to 1996. Yet, they later realized the limitations of their early interpretation and published multiple papers and reviews to correct their active site model [14].

The Crucial Role of Experimental Conditions

The core issue centers on the conditions of ex situ XPS experiments—specifically, whether:

i. the Zn/Cu(111) sample is cooled under vacuum after evacuating the CO_2/H_2 reaction gas at the reaction temperatures or.
ii. the Zn/Cu(111) sample is cooled in the presence of the reaction gas.

These two approaches yield distinctly different results.

Key Findings and the Current State of the Debate

In scenario (i), formate ($HCOO_a$), which is the most stable adsorbed species in the steady state on Zn/Cu(111), rapidly decomposes once the reaction gas is removed, leaving surface oxygen behind. This surface oxygen then oxidizes Zn to ZnO. In contrast, in scenario (ii), the sample is cooled while still exposed to the reaction gas, preventing decomposition of the formate species and thereby preserving the steady-state surface structure.

We highlighted this critical distinction in our correspondence to *Science*, but the debate regarding the true nature of the active sites continues [12].

Surface Structure of the Cu–Zn Alloy

Here, let us illustrate the structure of the Cu–Zn alloy surface. Figure 11.12 shows scanning tunneling microscope (STM) images of the clean Cu(111) surface and the Cu(111) surface after deposition of Zn atoms [14]. In the Zn/Cu(111) system at a coverage of $\Theta_{Zn} = 0.25$, the sample was heated to 560 K following Zn deposition at room temperature. In image (b), the bright spots correspond to Zn atoms, indicating that Zn atoms replace Cu atoms on the surface. Even upon heating, Zn does not diffuse into the bulk material. This suggests that the alloying occurs only at the surface.

This surface alloy stabilization is believed to arise from the fact that surface alloys are thermodynamically more stable than their bulk counterparts. Furthermore, partial charge transfer between Cu and Zn during alloy formation (for example, $Cu^{\delta-}$ and $Zn^{\delta+}$) likely contributes to stabilization through ionic interactions. Notably, the stability of the CuZn surface alloy also depends on the Zn composition and the underlying Cu surface structure. While Zn deposition significantly enhances catalytic activity on Cu(111), it does not exhibit similar effects on Cu(110) or Cu(100). Although these are all Cu surfaces, the catalytic activity varies depending on the specific crystallographic orientation.

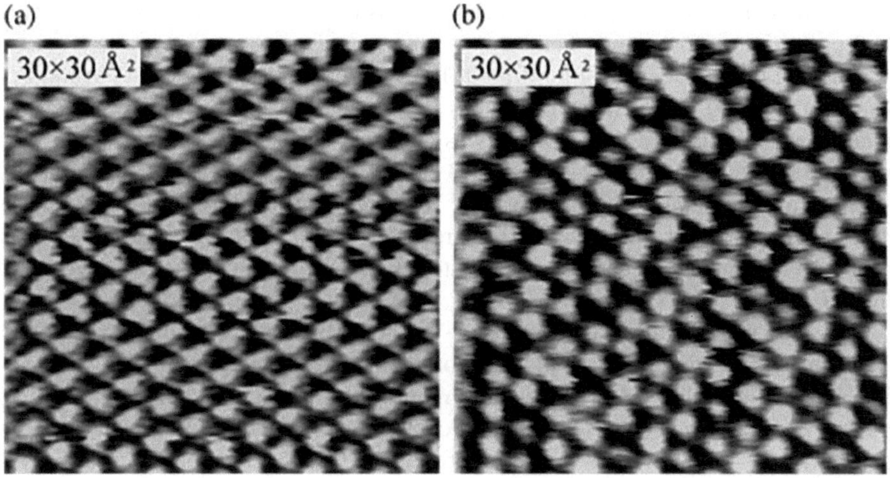

Fig. 11.12 STM images of (**a**) clean Cu(111) surface, (**b**) Zn/Cu(111) surface (after heating at 560 K) Zn coverage = 0.25. (Reprinted from [14] with the permission of Elsevier)

Complexity of Active Sites in Methanol Synthesis

Although the Zn/Cu(111) surface alloy has been identified as highly active for methanol synthesis, determining the precise location of the active site is not straightforward. In fact, even pure Cu surfaces show some catalytic activity for methanol production. When Zn is added, the formation of a CuZn alloy leads to a substantial enhancement of activity, but the key question remains: where in the reaction process does this enhancement occur?

To address this, it is necessary to identify the rate-determining step of the methanol synthesis reaction. The overall reaction process for methanol synthesis from CO_2 can be broadly divided into the following steps:

$$CO_2 \rightarrow HCOO_a \rightarrow CH_3O_a \rightarrow CH_3OH \tag{11.8}$$

Here, CO_2 is initially hydrogenated to formate ($HCOO_a$), then further hydrogenated to methoxy species (CH_3O_a), and finally converted to methanol. Understanding which step is accelerated by the CuZn alloy is essential for elucidating the true nature of the active site.

Role of Zn in the Hydrogenation Pathway

When investigating the effect of Zn on the initial formate formation process, it was found that there is no significant difference between Cu(111) and Zn/Cu(111). This indicates that formate is formed even without the presence of Zn and that Zn does not promote this initial step. However, as demonstrated below, Zn is essential for the hydrogenation of formate to methoxy.

Figure 11.13 presents in situ infrared (IR) absorption spectra of Cu(111) and Zn/Cu(111) (θ_{Zn} = 0.19) surfaces after CO_2 hydrogenation [7]. During the measurement, a gas mixture of CO_2:H_2 = 1:3 was introduced at a pressure of 760 Torr, and the substrate temperature was set to 343 K. On Cu(111), a peak attributed to the symmetric stretching vibration of the formate OCO group was observed at 1320 cm^{-1}. However, the asymmetric stretching vibration peak ν_a(OCO) was not observed on Cu(111) because its associated dipole moment change lies parallel to the surface.

On Zn/Cu(111), the ν_a(OCO) peak was clearly observed at approximately 1540 cm^{-1}, indicating that formate is adsorbed in a tilted configuration and suggesting that it is coordinated in a bidentate manner to the Zn–Cu site. This ν_a(OCO) peak was also detected on the Cu/ZnO catalyst, confirming that formate adsorbs across the Cu–Zn alloy sites. Furthermore, on Zn/Cu(111), a peak due to the C–O stretching vibration of methoxy emerged around 1080 cm^{-1}, providing experimental evidence that methoxy is indeed an intermediate. Collectively, these observations indicate that Zn plays a crucial role in promoting the hydrogenation of formate to methoxy.

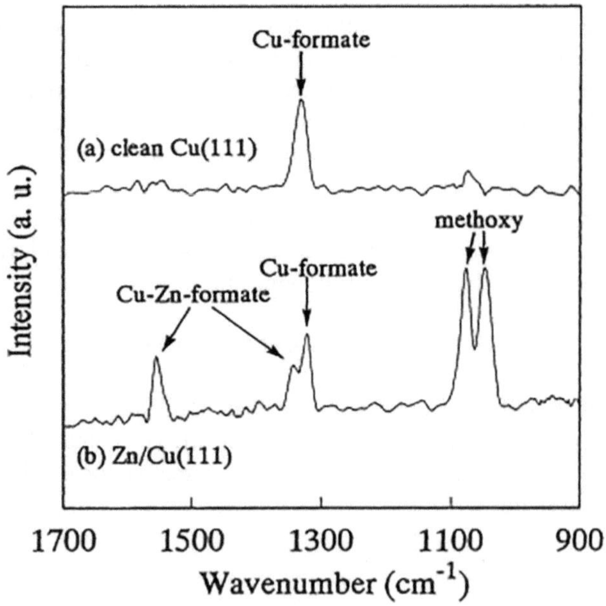

Fig. 11.13 In situ IR absorption spectra of formate species and methoxy species on clean Cu(111) and Zn/Cu(111) during CO_2 hydrogenation at 343 K and 760 torr. H_2/CO_2 = 1, 343 K, 1 atm total pressure. (Reprinted from [7] with the permission of Springer Nature)

Methanol Synthesis Activity of Cu Surfaces and Identification of Intermediates

Although the CuZn surface alloy significantly promotes methanol synthesis from CO_2, even pure Cu surfaces exhibit a small but measurable catalytic activity. In Fig. 11.10 ($\theta_{Zn} = 0$), the turnover frequency (TOF) is not zero, suggesting that pure Cu has some intrinsic activity. While the precise nature of the active site on pure Cu is not yet clear, experimental evidence suggests that defect sites, such as step edges, are responsible.

Nakamura and colleagues continue to investigate the mechanism of methanol synthesis, focusing on the identification of intermediates beyond formate. However, key intermediates such as dioxymethylene (CH_2COO_a) and formaldehyde ($HCHO_a$) are thermally unstable and decompose readily at room temperature. To stabilize and observe these intermediates on the surface, hydrogenation of formate ($HCOO_a$) is required. However, due to the relatively low H_2 dissociation capability of Cu, temperatures above 400 K are necessary for hydrogenation on Cu surfaces.

To overcome this, formate was pre-formed from formic acid on the Cu(111) surface ($HCOO_a$/Cu(111)), and then atomic hydrogen was reacted with this pre-formed formate while maintaining the surface at a low temperature (around 150 K). Atomic hydrogen was generated by passing H_2 molecules over a red-hot tungsten wire. Through these experiments, dioxymethylene (CH_2COO_a) and formaldehyde ($HCHO_a$) were successfully generated. Their formation was confirmed by infrared absorption spectroscopy (IRAS) and temperature-programmed desorption (TPD) experiments.

Potential Diagram for Methanol Synthesis on Cu(111)

Based on the thermal decomposition behavior of dioxymethylene and the desorption temperature of formaldehyde, a potential energy diagram for the reaction on Cu(111) was constructed, as shown in Fig. 11.14. The ability to construct such a diagram is a strong indication that the catalytic reaction mechanism has been elucidated. On the left side of the diagram, reactants (CO_2 and H_2) are depicted, with the assumption that oxygen atoms (O_a) are always present on the surface. Moving to the right, hydrogenation progresses through intermediates such as bidentate-$HCOO_a$ (with two oxygen atoms bonded to Cu) and monodentate-$HCOO_a$, followed by the formation of dioxymethylene, formaldehyde, methoxy, and finally methanol.

The diagram reveals that bidentate-$HCOO_a$ and methoxy are thermally stable, while dioxymethylene and formaldehyde are thermally unstable. The highest energy barrier, or "mountain peak," just before the formation of $H_2COO_a + 4H_a + O_a$ represents the activation energy for methanol synthesis. In this model catalyst system,

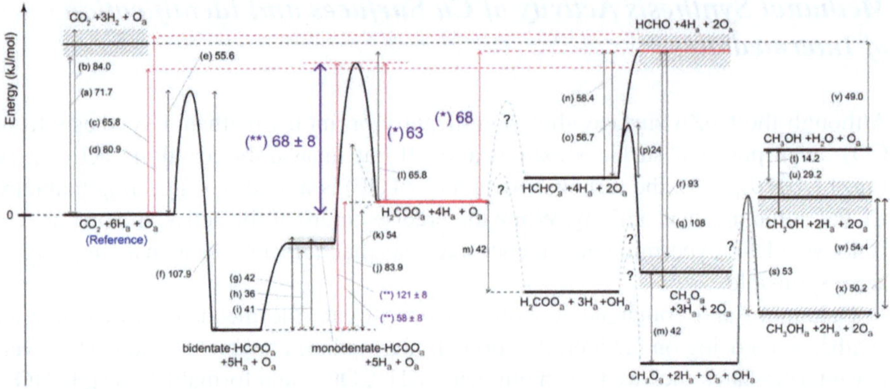

Fig. 11.14 Potential diagram for methanol synthesis by hydrogenation of CO_2 on a Cu(111) Surface. (Reprinted from [15] with the permission of American Chemical Society)

the apparent activation energy was determined to be 68 ± 8 kJ/mol, which aligns well with values for the Cu/ZnO powder catalyst, Zn/Cu(111), and Cu(111) (67–83 kJ/mol). Thus, the rate-determining step is currently understood to be the conversion of formate to dioxymethylene.

11.6 Activation of CO_2

Mechanistic Insights into Formate Formation from CO_2

Let us look at an interesting phenomenon in the formation of formate from CO_2. The initial reaction step for CO_2 on the Cu surface is the hydrogenation of CO_2 with adsorbed hydrogen atoms to form formate ($CO_2 + H_a \rightarrow HCOO_a$). Our research has revealed that this elementary process does not follow the usual Langmuir-Hinshelwood (LH) mechanism—where both reactants are adsorbed on the surface and then react—but rather follows the Eley-Rideal (ER) mechanism [16]. In the ER mechanism, a gas-phase molecule directly collides with and reacts with a surface-adsorbed species without undergoing prior adsorption. In this case, the energy of the gas-phase molecule contributes directly to overcoming the activation energy of the reaction.

During our analysis of the rate equation for formate formation in a model catalyst system, Nakamura hypothesized the possibility of the ER mechanism. To verify this, Nakamura et al. conducted molecular beam experiments with CO_2 to directly test the ER mechanism. In such experiments, proving a direct reaction involves showing that the reaction rate depends on the energy of the CO_2 molecule, not on the surface temperature.

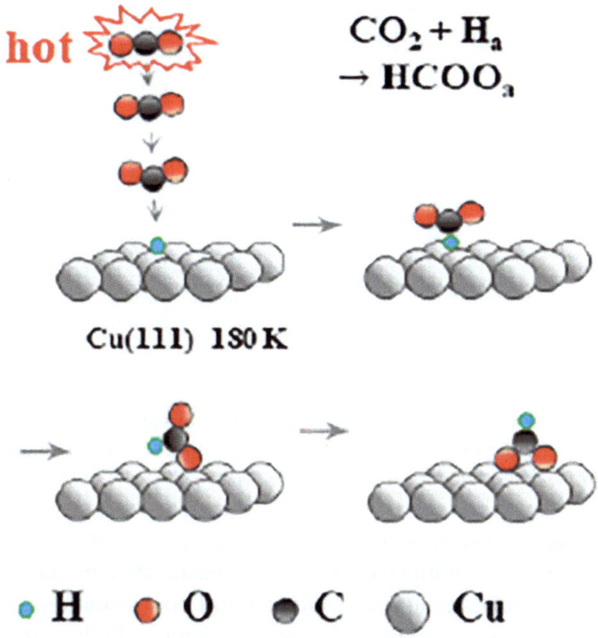

Fig. 11.15 CO_2 molecular beam experiment to prove the ER mechanism. A Cu surface on which atomic hydrogen has been pre-adsorbed is irradiated with the CO_2 molecular beam. The surface is cooled to about 180 K, while a large amount of thermal energy is applied to the CO_2

As shown in Fig. 11.15, Nakamura et al. adsorbed atomic hydrogen on the surface of Cu(111) (by the method described earlier) and then collided CO_2 molecules with controlled translational and vibrational energies onto this surface. The formation of formate was detected by infrared reflection absorption spectroscopy (IRAS) and temperature-programmed desorption (TPD), allowing us to measure the reaction probability (reaction rate).

Figure 11.16 illustrates the dependence of the reaction probability on (**a**) the surface temperature, (**b**) the translational energy of CO_2, and (**c**) the vibrational energy of CO_2. The results clearly show that the reaction probability does not depend on the surface temperature but strongly depends on the translational and vibrational energies of CO_2. Notably, when comparing the slopes ($\Delta \log P_0/\Delta E$), the slope associated with the vibrational energy is much steeper—by approximately eleven orders of magnitude—indicating that the vibrational energy of CO_2 is directly utilized for formate synthesis.

Further insights from density functional theory (DFT) calculations revealed that among the vibrational modes, changes in the O–C–O bond angle mode are especially influential. When the initially linear O–C–O axis (associated with sp-hybrid orbitals) bends, there is a significant change in the molecular orbital energy, and the

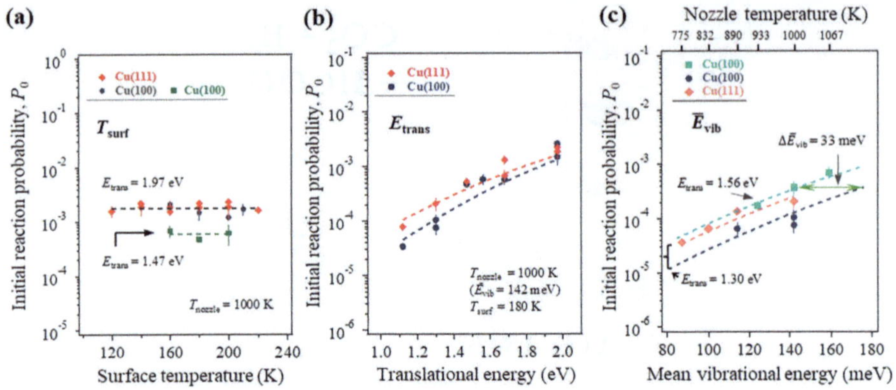

Fig. 11.16 Dependence of the reaction probability (P_0) of formate formation by the reaction of a CO_2 molecular beam with H/Cu(111) on surface temperature (**a**), translational energy (**b**), and vibrational energy (**c**). (Reprinted from [16] with the permission of Springer Nature)

energy level of the lowest unoccupied molecular orbital (LUMO) decreases. This makes CO_2 more reactive with electron-rich (or basic) species such as hydrides.

In the future, this fundamental understanding of how vibrational energy contributes to CO_2 activation could inform the development of high-efficiency processes for CO_2 utilization, especially in applications requiring advanced and high-cost catalysis.

References

1. Kylee H, Grim RG, Tao L (2021) A comparative techno-economic analysis of renewable methanol synthesis pathways from biomass and CO_2, Preprint edn. National Renewable Energy Laboratory, Golden, CO. NREL/CP-5100-78547
2. Cui P, Zhao L, Long Y, Dai L, Hu C (2023) Carbon-based Electrocatalysts for acidic oxygen reduction reaction. Angew Chem Int Ed 62:e202218269 (1–9)
3. Chinchen GC, Denny PJ, Parker DG, Spencer MS, Whan DA (1987) Mechanism of methanol synthesis from CO_2 /CO/H_2 mixtures over copper/zinc oxide/alumina catalysts: use of ^{14}C-labelled reactants. Appl Catal 30:333–338
4. Lee JS, Lee KH, Lee SY, Kim YG (1993) Comparative study of methanol synthesis from CO_2 /H_2 and CO/H_2 over a Cu/ZnO/Al_2O_3 catalyst. J Catal 144:414–424
5. Sahibzada M, Metcalfe IS, Chadwick D (1998) Methanol synthesis from CO/CO_2 /H_2 over Cu/ZnO/Al_2O_3 at differential and finite conversions. J Catal 174:111–118
6. Choi Y, Futagami K, Fujitani T, Nakamura J (2001) The difference in the active sites for CO_2 and CO hydrogenations on Cu/ZnO-based methanol synthesis catalysts. Catal Lett 73:27–31
7. Nakamura J, Choi Y, Fujitani T (2003) On the issue of the active site and the role of ZnO in Cu/ZnO methanol synthesis catalysts. Top Catal 22:277–285
8. Sheffer GR, King TS (1989) Potassium's promotional effect of unsupported copper catalysts for methanol synthesis. J Catal 115:376–387
9. Sheffer GR, King TS (1989) Differences in the promotional effect of the group IA elements on unsupported copper catalysts for carbon monoxide hydrogenation. J Catal 116:488–497

10. Fujitani T, Nakamura I, Uchijima T (1997) Nakamura J (1997). Surf Sci 383:285
11. Kattel S, Ramirez PJ, Chen JG, Rodriguez JA, Liu P (2017) Active sites for CO_2 hydrogenation to methanol on Cu/ZnO catalysts. Science 355:1296–1299
12. Nakamura J, Fujitani T, Kuld S, Helveg S, Chorkendorff I, Sehested J (2017) Comment on "active sites for CO_2 hydrogenation to methanol on Cu/ZnO catalysts". Science 357:eaan8074
13. Kattel S, Ramirez PJ, Chen JG, Rodriguez JA, Liu P (2017) Response to comment on "active sites for CO_2 hydrogenation to methanol on Cu/ZnO catalysts". Science 357:eaan8210
14. Fujitani T, Nakamura J (2000) The chemical modification seen in the Cu/ZnO methanol synthesis catalysts. Appl Catal A General 191:111–129
15. Takeyasu K, Sawaki Y, Imabayashi T, Marsha Putra SE, Halim HH, Quan J, Hamamoto Y, Hamada I, Morikawa Y, Kondo T, Fujitani T, Nakamura J (2022) Hydrogenation of Formate species using atomic hydrogen on a Cu(111) model catalyst. J Am Chem Soc 144:12158–12166
16. Quan J, Muttaqien F, Kondo T, Kozarashi T, Mogi T, Imabayashi T, Hamamoto Y, Inagaki K, Hamada I, Morikawa Y, Nakamura J (2019) Vibration-driven reaction of CO_2 on Cu surfaces via Eley–Rideal-type mechanism. Nat Chem 11:722–729
17. Choi Y, Futagami K, Fujitani T, Nakamura J (2001) The role of ZnO in Cu/ZnO methanol synthesis catalysts—morphology effect or active site model? Appl Catal A General 208:163–167

Chapter 12
Thinking Toward the Future

Abstract This final chapter highlights the critical roles of citizens, policy frameworks, and philosophical reflection in the pursuit of carbon neutrality. Active civic engagement, as seen in France's Climate Citizens' Assembly, can catalyze policy innovation and public support. Similarly, model cities like Sonderborg, Denmark, demonstrate how localized, community-driven efforts can serve as national and global exemplars. This chapter cautions against overreliance on emerging high-tech solutions and instead emphasizes reorganizing and deploying existing technologies that are accessible and effective. In this context, methanol emerges as a promising transitional energy carrier due to its compatibility with existing infrastructure and its potential for immediate emission reductions. Carbon pricing is introduced as a key economic mechanism to internalize the cost of carbon emissions and promote cleaner technologies. Beyond technical and policy tools, this chapter calls for philosophical engagement with the ethical dimensions of energy use, resource conservation, and long-term sustainability. Embracing circular economy principles and rethinking humanity's relationship with energy are essential to building a resilient, carbon-neutral future. The vision of a methanol-based society embodies these ideals, integrating practicality, innovation, and environmental responsibility.

Keywords Policy framework · Citizen engagement · Carbon pricing · Model cities · "Mottainai" spirit

12.1 Citizen Movements and Policy Decisions

The Role of Citizens in Shaping Carbon-Neutral Policies

To effectively promote carbon-neutral policies, political decision-making is essential. However, in many cases, policy decisions are driven primarily by economic considerations, resulting in policies that often fail to achieve meaningful CO_2 emissions reductions. In democratic societies, the voices of citizens should be central to discussions of carbon neutrality. Indeed, the active participation of citizens is likely

J. Nakamura, M. Ishida, *Blueprint for a Methanol Society*,
https://doi.org/10.1007/978-981-95-3465-4_12

to exert the greatest influence on both political and economic systems, since governments and companies ultimately exist to serve citizens.

A notable example of this is the Climate Citizens' Assembly in France, which has played a significant role in shaping national policy on climate change.

France's Climate Citizens' Assembly: A Model for Public Participation

In France, the 2019–2020 Citizens' Convention for Climate (Convention Citoyenne pour le Climat, CCC) brought together 150 randomly selected citizens who adopted 149 policy proposals [1]. President Macron committed to submit 146 proposals "without filter," and as of mid-2021, about 40% of these proposals had been incorporated into the Climate and Resilience Law. Key measures include expansion of renewable energy, stricter vehicle emissions standards, improved building energy efficiency, and plastic use restrictions. The French Citizens' Convention for Climate compensated its 150 participants at a rate equivalent to jury service—roughly €86 per day, totaling about €1462 per participant for the entire process. In addition, participants received an hourly wage of €10.03 if the sessions overlapped with work hours and had all travel, accommodation, meals, and childcare expenses fully covered. This comprehensive support ensured inclusive participation regardless of socioeconomic background.

Similar citizens' assemblies have since taken place in the UK and Germany, contributing to national policy discussions. In all these cases, collaboration with expert groups (e.g. technical and legal) was essential to ensure proposals were scientifically and legally sound.

While citizens' assemblies are influential in democracies, they are most effective when linked to expert analysis and legislative mechanisms, especially when promoting complex solutions like carbon capture or systemic efficiency improvements.

Expanding Citizen Participation across Europe

Following France's example, climate citizens' assemblies have also been held in the United Kingdom and Germany, helping to shape policy directions in those countries. This approach, which directly links citizens' voices to policy decisions, is a healthy model for inclusive governance.

The Need for Collaboration with Experts

Despite the promising role of citizen assemblies, there are limitations. In the proposals submitted by the French Citizens' Assembly, there are excellent ideas related to CCS (carbon capture and storage), renewable energy adoption, and transparency in CO_2 emissions. However, quantitative discussions about energy efficiency, sustainable social structures, and future visions appear to be lacking. This highlights the need for collaboration with academia and engineers to ensure that policy proposals are grounded in robust scientific and technical analysis.

Ultimately, while citizen engagement is indispensable, it must be complemented by open dialogue among diverse stakeholders, including experts and policymakers. In this context, the authors encourage consideration of the potential role of methanol-based energy systems in achieving a carbon-neutral society. Above all, the authors wish to express my respect and admiration for the pioneering efforts of citizens' assemblies in Europe.

12.2 Model City Planning

Practical Approaches to Methanol Integration in Urban Areas

When considering the introduction of methanol as an energy carrier in society, it is most practical to select a specific region and establish a distributed energy system. But what kind of regional plan should be developed? As a prerequisite, the integration of renewable energy is essential. A realistic first step involves utilizing electricity generated from solar cells, wind power, and hydropower as the main grid power source, while methanol serves as an adjustment power source. Deploying methanol fuel cells within the region would also be beneficial.

In these regional systems, methanol can be synthesized from CO_2 and H_2. CO_2 can be collected from emissions produced by thermal power plants and methanol fuel cells and transported to a methanol production plant via pipeline or trucks, depending on the local infrastructure and scale. H_2 can be generated by water electrolysis using renewable electricity. Methanol can be stored in tanks and supplied to local demand through pipelines. Existing gasoline stations could be repurposed to promote methanol-fueled vehicles. In industrial zones, methanol and hydrogen could be used as fuels, with hydrogen also produced by reforming methanol. Such a city or industrial zone would ideally combine centralized grid power with distributed energy systems.

The Importance of Heat Management

A critical aspect of regional planning is the effective use of waste heat. Discarding heat should be avoided wherever possible. For instance, it would be highly beneficial to utilize waste heat from thermal power plants for district heating or for the steam reforming reaction of methanol.

Case Study: Sonderborg, Denmark

The introduction of methanol use is particularly effective in small cities with strong commitments to carbon neutrality. Denmark, a country known for its progress in carbon neutrality, provides a notable example. The city of Sonderborg is steadily advancing its "ProjectZero" initiative, which aims to achieve carbon neutrality by 2030 [2]. As of August 2024, Sonderborg has reduced its CO_2 emissions by 66% compared to 2007. It is estimated that a further reduction of 240,000 tons is required to achieve complete carbon neutrality.

Sonderborg actively adopts renewable and decentralized energy systems, using smart grids to manage local energy consumption efficiently. The city leverages wind, solar, and biomass energy. Key features of this approach include the optimization of energy consumption, increased self-sufficiency in local energy supply, and the construction of energy-efficient buildings. Sonderborg hosts a concentration of small- and medium-sized manufacturing industries, particularly in machinery, metal processing, and electronics. When renewable energy generation is insufficient, natural gas-fired power generation serves as a backup source. While CO_2 capture is not yet implemented, it is reportedly under consideration. Natural gas also plays a role in the district heating system, providing heat to both residential and commercial users. However, there are plans to reduce natural gas use and further increase the share of renewable energy.

Smart City Heat Utilization

Denmark's smart city initiatives emphasize intensive heat utilization, and Sonderborg is no exception. The city employs efficient district heating and industrial heat energy use. Heat energy is supplied from centralized power plants and distributed to homes and businesses. The district heating system relies not only on fossil fuels but also on renewable sources such as solar thermal energy, biomass, and waste heat. Additionally, the introduction of heat pump technology is being actively pursued. Smart management systems have been implemented to optimize the balance between heat demand and supply. These efforts have significantly improved energy efficiency and contributed to the reduction of fossil fuel consumption and CO_2 emissions.

12.3 High-Tech Is Not Always Effective

Practical Innovation for Global Warming Mitigation

What constitutes true innovation in the context of climate change? Genuine innovation involves technologies that can be easily implemented not only in developed countries but also in developing countries. Technologies that are limited to developed regions will not solve the global warming crisis. High-cost, high-tech solutions cannot be introduced to society immediately—especially considering that global warming is an urgent issue. This discussion emphasizes that a methanol society is an approach that can be introduced quickly and practically.

However, in reality, particularly in developed countries, there is often a prevailing belief that advanced high-tech solutions alone will address global warming. It is argued here that methanol is a key substance for advancing climate change measures. As an energy carrier, methanol has been compared with liquid hydrogen and ammonia. Nevertheless, the authors do not claim that methanol is superior in every context, nor does it take a confrontational stance against other technologies. Depending on the region, liquid hydrogen, ammonia, or methane may be more suitable than methanol. It is important to preserve diversity in technological approaches and engage in broad, quantitative discussions, rather than competing solely based on scientific or technological superiority.

The Practical Value of "Low-Tech" Solutions

Methanol synthesis and utilization can be considered "mature and accessible technology" rather than cutting-edge or high-cost innovation. The following episode illustrates the practicality of methanol reforming through a real-world example. The author, Ishida, focused on methanol–water solution around 20 years ago. The reforming reactor shown in Fig. 12.1 was handmade in the laboratory. The device was quickly assembled to prepare reformed gas for purification experiments using a hydrogen storage alloy. A reforming catalyst, provided by a catalyst manufacturer, was packed into a stainless steel sample bottle and simply heated using an electric heater. The heater's temperature was regulated by a controller, and a methanol–water solution was injected with a pressure pump. Even with this simple setup, there were no issues producing reformed gas.

Interestingly, Ishida's original field was electrical engineering, and he had almost no prior experience constructing chemical reaction processes. The fact that such a

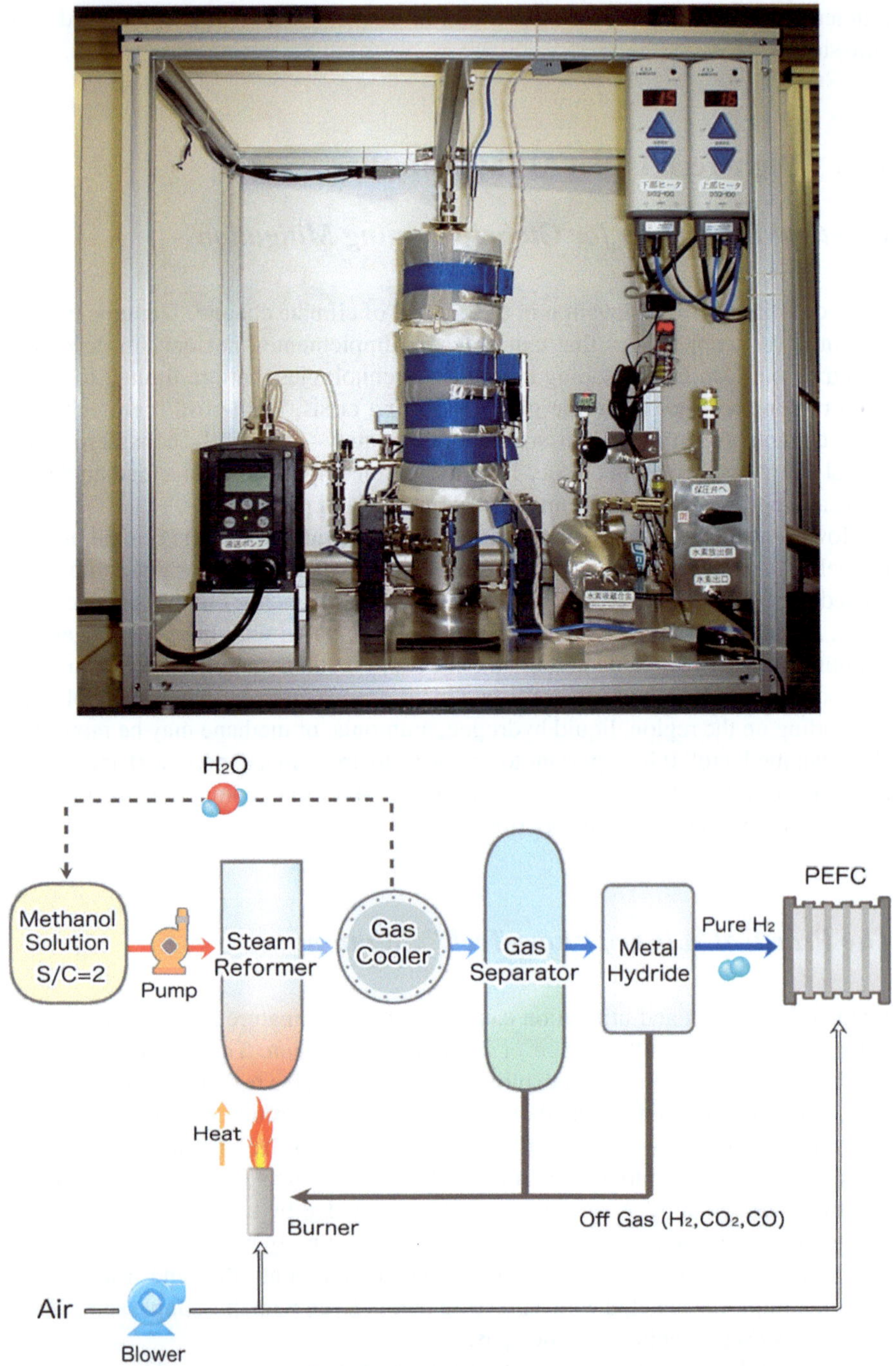

Fig. 12.1 (**a**) Homemade methanol reforming reactor and (**b**) a power generation system that combines a hydrogen storage device and a PEFC

system could be easily built by a novice underscored the practicality and reliability of methanol-based technology. In contrast, attempting steam reforming of methane would require temperatures exceeding 700 °C, significantly increasing the difficulty of implementation.

Reconsidering the Role of High-Tech in Energy and Climate Strategies

Conversely, utilizing energy from liquid hydrogen and ammonia is relatively high-tech. Although ammonia has long been used in chemical and agricultural fields, its application as an energy carrier is still new. The anticipation surrounding these carriers stems from the fact that they represent extremely challenging tasks: liquid hydrogen must be cooled to near absolute zero at −253 °C (20 K), and ammonia is a hazardous substance that poses difficulties in handling. While these challenges lower the likelihood of immediate implementation, the potential impact if they are overcome is considerable.

In contrast, methanol presents a simpler technological pathway, with the primary challenge lying in the source of CO_2 rather than in the conversion reaction itself. Consequently, methanol's implementation may seem straightforward, lacking the allure of overcoming an extreme technological challenge. This perception often leads to an underestimation of methanol's value—a pitfall shaped by modern biases, particularly prevalent in universities, research institutes, and companies that consider themselves advanced.

The Strength of Low-Tech Solutions and Practical Priorities

While such a mindset may be effective in the rapidly evolving digital and information technology fields, it does not translate well to mature sectors like energy. In these fields, even if high-tech solutions are realized, they can lead to increased resource consumption without a commensurate positive impact—sometimes even a negative one. In contrast, the so-called low-tech solutions that emphasize early implementation, economic efficiency, and responsible use of shared, finite resources can have far-reaching effects. This book encourages a perspective that values these qualities.

Although high-tech approaches may initially appear more powerful, they often prove to be more fragile in practice. In contrast, low-tech solutions—while seemingly modest—tend to offer greater resilience and reliability. This contrast extends far beyond the choice of energy carriers and can be observed across a range of societal issues: from "renewable energy vs. fossil fuels" and "bicycles vs. flying cars" to "academic freedom vs. centralization," "small- and medium-sized enterprises vs.

large corporations," "rural vs. urban development," and even "democracy vs. monarchy." In each of these cases, the central challenge lies in identifying and reinforcing resilient approaches that align more closely with long-term sustainability and human well-being.

Given these considerations, it is wiser to harness the abundant resources provided by nature and avoid gambling on risky high-tech solutions. The key boundary conditions are:

i. Immediate realization of effective outcomes
ii. Incorporation of modest behavioral changes
iii. Regional self-sufficiency
iv. Business sustainability
v. Global applicability

The Practical Value of Methanol and Supporting Infrastructures

Methanol carriers, lightweight and compact mobility solutions, the development of fossil fuel-free model towns (aligned with the holonism concept), and the introduction of underground pumped storage power generation with flood control are essential. Above all, behavioral changes to minimize energy use are crucial.

Underground Pumped Storage Power Generation

Underground pumped storage power generation, as depicted in Fig. 12.2, is a vital form of energy storage and supply that plays a critical role in stabilizing the electricity grid and supporting the growth of renewable energy. The basic principle involves pumping water from a lower reservoir to an upper reservoir during periods of surplus electricity. True to its name, the upper reservoir is typically located underground. When electricity demand surges and there is a shortage of supply, water is released from the upper reservoir, flowing through turbines to generate electricity as it descends under the force of gravity.

One of the primary advantages of underground pumped storage is its ability to store large volumes of energy over long periods, effectively balancing out the fluctuations inherent in renewable sources like solar and wind. Additionally, it can rapidly supply electricity during peak demand, thereby enhancing the stability of the power grid. This system also enables the efficient use of surplus electricity, reducing overall energy waste. From the perspective of disaster resilience, placing these systems underground improves their durability against natural hazards such as earthquakes and typhoons. Furthermore, it allows for the effective use of underground

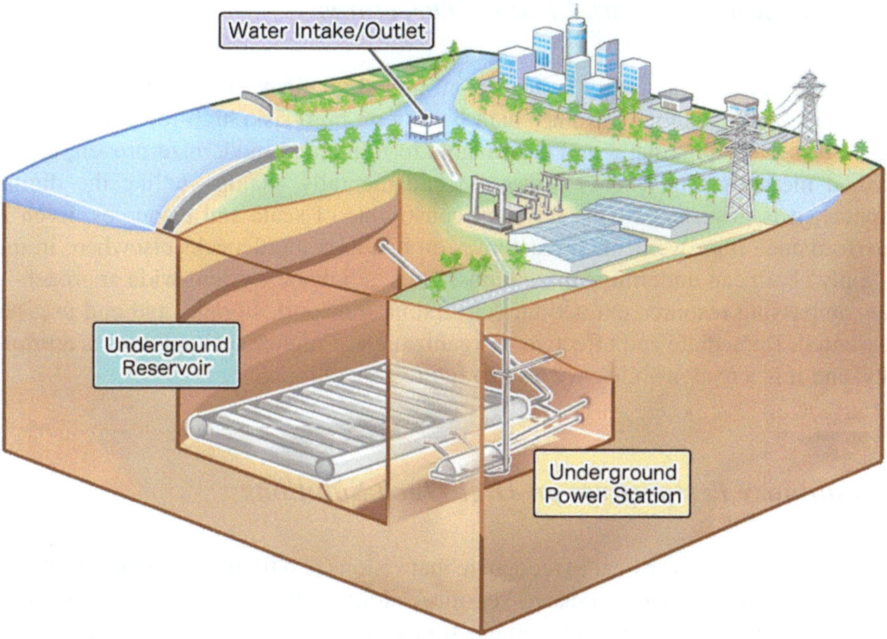

Fig. 12.2 Image illustration of a disaster response type deep underground pumped storage power generation facility

spaces in urban areas, preserving valuable above-ground land for other purposes. This approach can also enhance regional energy self-sufficiency and security.

However, underground pumped storage presents significant challenges. The initial investment is substantial, reflecting the high costs of constructing underground reservoirs and power generation infrastructure. Large-scale civil engineering efforts are required, necessitating careful environmental impact mitigation. Moreover, successful deployment depends on favorable geographical conditions, such as sufficient elevation differences. These factors must be thoroughly evaluated when planning new installations.

Figure 12.2 illustrates a deep underground pumped storage facility designed to also serve as a disaster mitigation measure. Although it has been implemented as a large-scale energy storage technology, the feasibility of deep civil engineering still requires further validation. Such systems can be strategically placed near rivers and estuaries vulnerable to flooding. By emptying the underground reservoir in advance of extreme rainfall events, the facility can function as an emergency waterway, helping to avert large-scale economic losses from flooding. Under normal circumstances, it adjusts power supply and demand within the grid, offering significant economic advantages and flexibility in power source selection.

Broader Reflections and Future Directions

Achieving carbon neutrality is an urgent, global challenge that demands realistic and sustainable solutions. Rather than betting on high-risk, high-reward technologies, it is more pragmatic and dependable to refine and modernize proven, established methods. The energy sector is fundamentally analog—unlike the digital sphere, it is governed by the physical constraints of scale and resources. Even if performance improvements are achieved in one area, bottlenecks elsewhere in the supply chain can undermine these gains. Thus, a holistic, system-wide approach—encompassing resource extraction, conversion, transport, storage, and end use—is essential. Immediate and effective reorganization of mature technologies is a priority, and it is a responsibility we owe to future generations.

Cautionary Perspectives on High-Tech Solutions

At this juncture, it is critical to recognize that visions of a future dominated by flying cars, space travel, and pervasive robotics, along with unchecked digitalization, involve inherently high energy and electricity consumption. While these technologies may seem promising, increasing reliance on them can drain human vitality and introduce new vulnerabilities. In contrast, energy is a gift from nature—a shared resource that should not be monopolized by a select few but responsibly managed by all of humanity.

These priorities are not merely technical—they are deeply ethical and social. The path to sustainability must be grounded in values such as fairness, mutual support, and the responsible use of shared resources. This perspective resonates with the philosophy of holonism, a systems-oriented approach that emphasizes interconnectedness and balance across all levels of society.

Holonism, embodied in the phrase "One for All, All for One," underscores the value of equitable, collective responsibility in energy use as we pursue carbon neutrality. Instead of centralizing authority or depending on a select few to drive innovation, holonism promotes distributed action and community-level cooperation. It encourages each part (or "holon") to function both independently and as a contributing unit to the whole—offering a powerful framework for addressing global challenges through locally grounded, scalable, and participatory efforts.

It is often said that "the very forces that bring prosperity to civilization also sow the seeds of its collapse." With careful wisdom and humility, a soft landing is possible, but unbridled desire inevitably leads to decline. While science and technology have certainly brought convenience, they have not always delivered a deeper sense of well-being; indeed, much of what once brought peace of mind has been lost. Therefore, it is essential not only to focus on the potential benefits of new technologies but also to remain vigilant about their possible downsides. Such balanced thinking is key to leaving a legacy of thoughtful stewardship and harmonious coexistence with the natural world for future generations.

12.4 The "Mottainai" Spirit

A defining feature of a methanol-based energy society is its emphasis on circulating carbon resources—essentially, closing the carbon cycle and achieving high overall energy efficiency. At the heart of this concept lies the idea of minimizing waste and making careful use of materials and energy. In Japan, this philosophy is encapsulated by the term "Mottainai," a spirit that can serve as a guiding principle for energy use. Wangari Muta Maathai of Kenya, shown in Fig. 12.3, is a Nobel Peace Prize-winning environmental activist who sympathized with the "Mottainai" spirit in her global environmental protection activities and worked to spread it internationally.

The "Mottainai" spirit represents a traditional Japanese value of avoiding waste and using resources prudently. It encourages recognizing the true worth of things and avoiding unnecessary disposal or waste, including reusing items whenever possible. This approach naturally leads to more effective resource use and extends to practices like minimizing food waste. By cherishing what we have, we also cultivate a sense of gratitude toward the people who created these items and to nature itself.

During Japan's Edo period (1603–1868), the "Mottainai" spirit permeated daily life. With limited resources, everything was reused. For example, since Japanese paper was expensive, used paper was recycled to make new paper. Kimonos and other textiles were carefully repaired and reused until they could no longer be worn, then repurposed as rags. Even metal items such as nails, pots, and tools were repeatedly repaired; broken metal products were melted down to create new tools. Figure 12.4 is a picture drawn in the Morisada Manko that illustrates a merchant collecting old umbrellas for repair and reuse, reflecting the widespread practice of resource conservation. The Morisada Manko is a kind of encyclopedia (35 volumes in total) that explains the customs and things of the late Edo period (1837–1868). The author is Kitagawa Morisada.

Extending "Mottainai" to Energy Use

The "Mottainai" spirit also applies to energy use and the circular flow of carbon resources. In this context, efficient heat utilization is particularly crucial. As already discussed, today's coal-fired power plants convert only about 50% of the energy in

Fig. 12.3 Nobel Peace Prize Laureate, Wangari Muta Maathai [3]. Reprinted from https://commons.wikimedia.org/wiki/File:Wangari_Maathai_in_2001.jpg under Creative Commons Attribution-ShareAlike 2.0 International License via Wikimedia Commons

coal into electricity (and older plants just 35%), with the remaining energy released as waste heat through turbine generators and lost to the sea via cooling water. This valuable heat could be harnessed for district heating, hot water supply in factories and buildings, or industrial processes, but in most cases, it goes unused. Although some factories have recently begun tapping into waste heat to improve energy efficiency, these efforts remain rare overall. As of 2024, 80% of CO_2 emissions originate from fossil resources, mostly from electricity generation, industry, and transportation—highlighting the growing importance of heat management in the future.

The methanol energy society aligns with the "Mottainai" spirit by making full use of energy in multiple forms. In a society where power from renewable energy sources is converted into methanol as an energy carrier, the inevitable energy conversion losses can be repurposed as heat for air conditioning, hot water supply, and industrial processes. Methanol, when used as a distributed power source, can also contribute to the development of smart communities. Beyond energy use, methanol can serve as a feedstock for plastics production and, after use, be recycled as a carbon or hydrogen source—closing the loop and minimizing waste. This model of

Fig. 12.4 Edo period umbrella recycling business [4]. Reprinted from https://dl.ndl.go.jp/pid/2592395/1/13 Digital Correction of National Diet Library, Japan

carbon and energy circulation creates a sustainable and resource-efficient society that truly embodies the "Mottainai" spirit.

Quantitative Advantages and Future Vision

To help communicate this vision of a future methanol-based energy society to citizens and stakeholders, it is important to present quantitative data and concrete examples:

i. **Improved energy efficiency:** Using methanol as an energy carrier enables not only the circulation of CO_2 but also increases the efficiency of renewable energy utilization. By capturing waste heat, the total energy efficiency can improve by more than 30% compared to direct use of renewable electricity (see Sect. 6.3).
ii. **Realizing the carbon cycle:** By cycling methanol through plastic production and recycling, carbon resources can be reused, potentially reducing global CO_2 emissions by billions of tons annually (see Sect. 7.2).
iii. **Local economic contributions:** Methanol as a distributed energy source enhances regional energy self-sufficiency and helps lower energy costs, benefiting local economies.

In summary, the methanol energy society represents a sustainable, efficient, and practical future for energy use—rooted in the "Mottainai" spirit of cherishing and effectively using every resource. It is crucial to carry forward this ethos and work together to realize a future that harmonizes human prosperity with environmental stewardship.

12.5 Carbon Pricing

Overview of Carbon Pricing

Carbon pricing is a policy instrument designed to provide incentives for companies and individuals to reduce their greenhouse gas (GHG) emissions by assigning a monetary cost to those emissions. The primary objective of carbon pricing is to internalize the economic cost of greenhouse gas emissions, thereby encouraging more environmentally responsible behavior.

Types of Carbon Pricing

There are two main approaches to carbon pricing:

i. **Carbon Tax**

   A carbon tax puts a price on the carbon emissions from burning fossil fuels such as coal, oil and natural gas. It is typically levied on the carbon dioxide (CO_2) content of these fuels, so the tax paid is proportional to the amount of CO_2 released when the fuel is burned. Table 12.1 compares the nominal tax rates at the time of introduction and the 2024 rates for a number of jurisdictions. It should be noted that there is no European Union-wide carbon tax; the EU operates an emissions trading system (EU ETS) instead, while individual member states such as Sweden, Finland and France levy their own carbon taxes. Canada introduced a federal carbon price in 2019 and its standard rate will rise from US$15 per tonne in 2019 to US$50–60 per tonne in 2024.

Table 12.1 Effects of carbon pricing in Sweden, the EU, Canada, Japan, and the USA

Country/ region	Introduction year	Initial price (USD/ton)	Current price (2024) (USD/ton)	CO_2 reduction (%)	GDP impact (%)
Sweden	1991	26–27	127	27	75
EU ETS	2005	10–15	70	35	30
Canada	2019	15	50–60	10	2
Japan	2021	2 (9–18)	2 + (9–18)	5	1
USA (nationwide)	N/A	N/A	N/A 30–40 in California state	N/A	N/A
China ETS	2021	7	12–15	N/A	N/A

Japan's "Global Warming Countermeasure Tax" began in 2012 and has been set at ~US$2 per tonne of CO_2 since 2016. The carbon tax in Japan is implemented as a surcharge on the existing Petroleum and Coal Tax. Introduced in 2012 as part of the "Tax for Climate Change Mitigation," this carbon tax imposes an additional levy based on the CO_2 emissions associated with fossil fuel consumption. As of 2025, the total tax burden—including both the base Petroleum and Coal Tax and the carbon surcharge—is approximately:

- US$9.00 per ton of CO_2 from coal.
- US$13.00 per ton of CO_2 from natural gas (LNG).
- US$18.00 per ton of CO_2 from oil products.

These rates are derived from energy-specific tax rates applied per unit of fuel (e.g., per kiloliter or per ton) and their corresponding CO_2 emission factors. Notably, coal—despite having the highest CO_2 emissions per unit energy—bears the lowest tax burden per ton of CO_2, which may weaken incentives for transitioning away from coal-fired power generation. Compared to many European countries, where carbon taxes or emissions trading systems often exceed US$50 to US$100 per ton of CO_2, Japan's effective carbon price remains relatively low. Furthermore, the disparity in tax burdens across fuel types raises concerns about the consistency and fairness of Japan's carbon pricing system. International organizations such as the OECD have pointed out that aligning tax rates more closely with carbon intensity would enhance the environmental effectiveness of Japan's climate policy.

The United States does not have a nationwide carbon tax. However, fossil fuel use is indirectly taxed through federal fuel taxes, which amount to approximately US$20–25 per ton of CO_2 for gasoline and diesel. Additionally, some states such as California operate cap-and-trade systems, where the market price of CO_2 allowances has ranged between US$30 and US$40 per ton in recent years. Although the federal government uses a "Social Cost of Carbon" of US$51 per ton in regulatory impact assessments, this figure does not translate into actual carbon pricing in the market.

China does not currently have a nationwide carbon tax, but it operates the world's largest national emissions trading system (ETS), which began full operation in 2021. The ETS initially covers only the power sector—particularly coal-fired power plants—requiring companies to hold allowances for their CO_2 emissions. The current carbon price in the Chinese ETS is approximately US$12–15 per ton of CO_2. Although the system does not yet cover heavy industries such as steel and cement, expansion to these sectors is planned. Thus, while Chinese companies do not pay a carbon tax, a carbon price is already embedded in certain sectors through market-based mechanisms.

Sweden introduced a carbon tax in 1991 at a rate of US$26–27 per tonne of CO_2, which was roughly US$26–27 at the time. The rate has been increased several times and, by 2023, reached about US$127 per tonne, one of the highest carbon tax rates in the world. Sweden's economic and environmental performance provides an illustrative case: real GDP grew roughly 75 percent from

1990 to 2016 while domestic CO_2-equivalent emissions fell about 26 percent. More recent data show that by 2021 Sweden's GDP had expanded about 92 percent and emissions had dropped about 33 percent compared with 1990 levels, and between 1990 and 2018 greenhouse gas emissions declined around 27 percent. These figures highlight that significant emissions reductions have been achieved alongside continued economic growth.

ii. **Cap-and-Trade System**

Under a cap-and-trade system, the government sets an overall emissions limit (cap) and allocates or sells emission allowances to companies within that limit. Companies must hold sufficient allowances to cover their emissions; if they fall short, they must purchase additional allowances on the market. This framework incentivizes firms to reduce emissions cost-effectively, since unused allowances can be banked or sold. The European Union Emissions Trading System (EU ETS), launched in 2005, covers over 10,400 industrial plants and power stations and around 350 airlines across the EU, Iceland, Norway, and Liechtenstein. In 2021, the cap was set at about 1.57 billion EU allowances (EUAs), each representing one tonne of CO_2 equivalent. Allowance prices climbed rapidly in recent years: the benchmark price passed €50/tCO_2 for the first time in May 2021, and the average auction price in 2024 was around €65–70/tCO_2 (≈US$70)—considerably higher than the "€50" figure cited earlier. Between 2005 and 2019, emissions from regulated facilities fell by roughly 35%, indicating the system's effectiveness, though free allocations and surplus permits in the early years tempered the incentive to invest in cleaner technologies.

Advantages and Challenges of Carbon Pricing

Carbon pricing offers several advantages:

- **Economic Efficiency:** By putting a price on carbon, market-based mechanisms encourage the least-cost emissions reductions.
- **Incentives for Emission Reduction:** Companies and consumers are encouraged to adopt cleaner technologies and practices.
- **Secondary Benefits:** Revenues generated from carbon pricing can be reinvested in environmental initiatives and social programs.

However, there are challenges to be addressed:

- **Setting the Right Price:** Prices that are too low may fail to reduce emissions effectively, while prices that are too high can burden the economy.
- **International Competitiveness and Carbon Leakage:** Differences in national systems can create competitive imbalances and shift emissions to countries with laxer regulations.
- **Equity Considerations:** Ensuring fairness and mitigating impacts on low-income households is essential.

Global Adoption and Economic Impact

The World Bank's 2023 update notes that 73 carbon pricing instruments (carbon taxes and emissions trading systems) were in force worldwide, covering about 23% of global GHG emissions.

The IPCC Special Report on 1.5 °C states that to limit global warming to 1.5 °C, global greenhouse gas emissions must be reduced by roughly 45% relative to 2010 levels by 2030. Achieving this target requires carbon prices that vary widely depending on model assumptions: integrated assessment models estimate values ranging from US$135 to US$6050 per tonne of CO_2e in 2030. Because these figures are sensitive to technological and policy conditions, they should be interpreted as broad indicators rather than precise forecasts. Separately, the High-Level Commission on Carbon Prices recommends carbon prices of around US$50–100 per tonne of CO_2 by 2030 to meet the Paris Agreement goals in an economically efficient way. The price range of US$135–550 per tonne cited earlier does not align with either the IPCC's broad cost estimates or the commission's recommendations.

In conclusion, international experience shows that well-designed carbon taxes and emissions trading systems can effectively reduce greenhouse gas emissions. When carbon prices are set at appropriate levels and accompanied by supportive policies, these instruments spur technological innovation, improve energy efficiency, and become central pillars of climate change mitigation. At the same time, policymakers must address key challenges such as setting the right price, protecting competitiveness, and ensuring social equity. Ongoing adjustments based on the latest data and scientific insights are essential to maintain the credibility and effectiveness of carbon pricing as part of a comprehensive sustainable development strategy.

References

1. home page of The Citizen's Convention on Climate. https://www.conventioncitoyennepour-leclimat.fr/en/
2. home page of Project Zero in Sonderborg, https://projectzero.dk/en/city-climate-action/
3. Photo by Kingkongphoto & Celebrity-photos.com (© 2005), CC BY-SA 2.0 via Wikimedia Commons
4. "Shuzen Manga" (Volume 6) p13, National Diet Library Digital Collection

Afterword

Humanity now stands at the precipice of an unprecedented crisis: climate change. This challenge surpasses the COVID-19 pandemic in terms of its long-term impact and global scope. Indeed, the pandemic has served as a harsh lesson, revealing the scale and complexity of systemic global crises. People around the world experienced restrictions on movement and daily life, yet despite this, public urgency regarding climate change remains alarmingly low.

Will we continue to squander energy, degrade the environment, and ultimately shorten our own lifespans? This pervasive lack of crisis awareness stems from a fundamental deficit of accurate knowledge and rigorous discussion. In this regard, scientists and engineers bear a profound responsibility. Without their expertise and leadership, the public cannot fully grasp the seriousness of the situation nor understand how to respond effectively.

This book has explored the imperative of creating artificial fuels, particularly methanol, as a cornerstone of a sustainable energy future. It is increasingly evident that humanity must harness solar energy to produce synthetic fuels through chemical processes. However, this marks a radical departure from the past, where chemical reactions have predominantly relied on fossil fuels for both energy and feedstocks.

This transformation presents formidable challenges. Historically, society has benefited from abundant and inexpensive fossil-derived energy. There was even a time when gasoline cost less than bottled water. Transitioning to synthetic fuels will likely require accepting significantly higher energy costs—possibly up to three times the price of conventional fuels. If we fail to make this shift, we risk exacerbating the climate crisis.

As a first step, we must recognize the scale of this challenge. Then, we must carefully assess and select the most appropriate synthetic fuel. **At the heart of the carbon neutrality debate lies a crucial question: which energy carrier should humanity adopt for the future?**

J. Nakamura, M. Ishida, *Blueprint for a Methanol Society*,
https://doi.org/10.1007/978-981-95-3465-4

This book argues that methanol is the most practical and promising candidate. It facilitates CO_2 capture and reuse, effectively closing the carbon cycle, and enables rapid deployment using existing infrastructure and technologies. This is not to dismiss other options—such as hydrogen, ammonia, or methane—all of which may play important roles depending on regional and sectoral needs. Yet, what remains missing is a comprehensive, quantitative comparison among these options.

Too often, public discussion about energy solutions focuses narrowly on switching to electric vehicles. But this alone will not solve the broader CO_2 problem. Power generation, the steel industry, the cement industry, and the chemical industry together account for about 65% of global CO_2 emissions. At the very least, we must outline a clear decarbonization roadmap for these key sectors.

The authors hope that this book can serve as a catalyst for such critical discussions, fostering the collective understanding and action required to build a truly sustainable future.

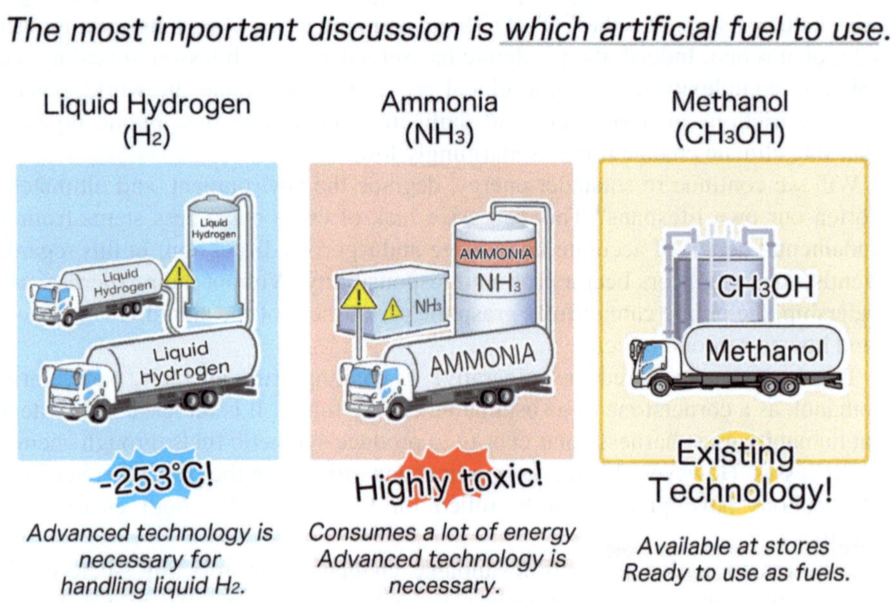

MIX
Papier aus verantwortungsvollen Quellen
Paper from responsible sources
FSC® C105338
FSC
www.fsc.org

If you have any concerns about our products,
you can contact us on
ProductSafety@springernature.com

In case Publisher is established outside the EU,
the EU authorized representative is:
Springer Nature Customer Service Center GmbH
Europaplatz 3, 69115 Heidelberg, Germany

Printed by Libri Plureos GmbH
in Hamburg, Germany